The Handbook of Infrared and Raman Spectra of Inorganic Compounds
and Organic Salts (a 4-volume set)

VOLUME 2

INFRARED AND RAMAN SPECTRAL ATLAS
OF INORGANIC COMPOUNDS AND ORGANIC SALTS:
RAMAN SPECTRA

The Handbook of Infrared and Raman Spectra
of Inorganic Compounds and Organic Salts
(a 4-volume set)

Volume 2
Infrared and Raman Spectral Atlas of Inorganic Compounds and Organic Salts:
RAMAN SPECTRA

Richard A. Nyquist, Curtis L. Putzig, and M. Anne Leugers

The Dow Chemical Company
Midland, Michigan

ACADEMIC PRESS
San Diego London Boston
New York Sydney Tokyo Toronto

ACADEMIC PRESS
525 B Street, Suite 1900, San Diego, CA 92101-4495, USA
1300 Boylston Street, Chestnut Hill, Massachusetts 02167
http://www.apnet.com

ACADEMIC PRESS LIMITED
24–28 Oval Road, London NW1   7DX, UK
http://www.hbuk.co.uk/ap/

LIBRARY OF CONGRESS CATALOGING-IN-PUBLICATION DATA

Nyquist, Richard A.
   Infrared and Raman spectral atlas of inorganic compounds
and organic salts / Richard A. Nyquist, Curtis L. Putzig, M.
Anne Leugers.
       p.   cm.
   Includes bibliographical references and indexes.
   Contents: v. 1. Text and explanations — v. 2. Raman spectra
— v. 3. Infrared spectra
   ISBN 0-12-523445-7 (v. 1). — ISBN 0-12-523446-5 (v. 2). —
ISBN 0-12-523447-3 (v. 3)
   1. Infrared spectra—Tables.   2. Raman effect—Tables.
3. Inorganic compounds—Spectra—Tables.   4. Organic
compounds—Spectra—Tables.   I. Putzig, Curtis L.
II. Leugers, M. Anne.   III. Title.
QC457.N927   1996
543'.08583—dc20   96-22175
                                                          CIP

PRINTED IN THE UNITED STATES OF AMERICA
96 97 98 99 EB 9 8 7 6 5 4 3 2 1

CONTENTS

Raman Spectra

The Handbook of Infrared and Raman Spectra of Inorganic Compounds
and Organic Salts (a 4-volume set)

VOLUME 2

INFRARED AND RAMAN SPECTRAL ATLAS
OF INORGANIC COMPOUNDS AND ORGANIC SALTS:
RAMAN SPECTRA

A. NUMERICAL INDEX OF RAMAN (R) AND INFRARED (IR) SPECTRA OF IONIC INORGANIC COMPOUNDS

Spectrum Number	Compound	Formula	Spectra IR	Spectra R	Vol. 4 IR
1	Boric acid	H_3BO_3	X	X	16
2	Lithium tetraborate	$Li_2B_4O_7$	X		18
3	Sodium tetraborate	$Na_2B_4O_7 \cdot 5H_2O$	X	X	20
4	Potassium tetraborate	$K_2B_4O_7 \cdot 8H_2O$	X	X	22
5	Sodium cyanide	NaCN		X	—
6	Potassium cyanide	KCN		X	—
7	Nickel (II) cyanide	$Ni(CN)_2 \cdot 4H_2O$	X	X	27
8	Copper (I) cyanide	CuCN	X	X	28
9	Zinc cyanide	$Zn(CN)_2$	X	X	29
10	Silver cyanide	AgCN	X		30
11	Platinous cyanide	$Pt(CN)_2 \cdot xH_2O$	X	X	31
12	Mercury (II) cyanide	$Hg(CN)_2$		X	32
13	Potassium cyanoargenate	$KAg(CN)_2$	X	X	33
14	Potassium cyanocuprate	$KCu(CN)_2 \cdot xH_2O$	X	X	34
14a	Potassium ferricyanide	$K_3Fe(CN)_6$		X	35
15	Sodium nitroferricyanide	$Na_2Fe(CN)_5NO_2 \cdot 2H_2O$		X	36
16	Sodium ferrocyanide	$Na_4Fe(CN)_6$		X	37
17	Potassium ferrocyanide	$K_4Fe(CN)_6$	X	X (X)	38
18	Potassium calcium ferrocyanide	$K_2CaFe(CN)_6$	X	X	39
19	Calcium ferrocyanide	$Ca_2Fe(CN)_6$		X	—
20	Potassium copper (II) ferrocyanide	$K_2Cu(CN)_6$		X	40
21	Lead ferrocyanide	$Pb_2Fe(CN)_6 \cdot xH_2O$	X	X	41
22	Iron (III) ferrocyanide	$Fe_4[Fe(CN)_6]_3$	X		42
23	Sodium cyanate	NaOCN	X	X	44
24	Silver cyanate	AgOCN	X		45
24a	Potassium thiocyanate	KSCN		(X)	46
25	Lead thiocyanate	$Pb(SCN)_2$	X	X	47
26	Iron (II) thiocyanate	$Fe(SCN)_2 \cdot 3H_2O$	X		48
27	Copper (I) thiocyanate	CuSCN	X	X	49
28	Silver thiocyanate	AgSCN	X	X	50

(X) in water solution
Multiple Xs denotes more than one spectra

Spectrum Number	Compound	Formula	Spectra IR	R		Vol. 4 IR
29	Sodium bicarbonate	$NaHCO_3$	X	X	(X)	54
30	Potassium bicarbonate	$KHCO_3$	X	X		55
31	Lithium carbonate	Li_2CO_3	X	X		56
32	Sodium carbonate	Na_2CO_3	X	X	(X)	57
32a	Potassium carbonate	K_2CO_3	X		(X)	
33	Calcium carbonate (calcite)	$CaCO_3$		X		61
34	Barium carbonate	$BaCO_3$		X		63
35	Lead carbonate	$PbCO_3$	X	X		64
36	Cadmium carbonate	$CdCO_3$	X	X		68
37	Guanidinium carbonate	$[(H_2N_2)C=NH_2]_2CO_3$	X	X		—
38	Barium thiocarbonate	$BaCS_3$	X	X		75
39	Magnesium silicide	Mg_2Si	X			76
40	Calcium silicide	Ca_2Si	X			77
41	Boron silicide	B_6Si	X			78
42	Titanium silicide	$TiSi_2$	X			79
43	Vanadium silicide	VSi_2	X			80
44	Manganese silicide	$MnSi_2$	X			81
45	Molybdenum silicide	$MoSi_2$	X			85
46	Cobalt orthosilicate	$CoSiO_4 \cdot xH_2O$	X			—
47	Barium zirconium silicate	$BaZrSiO_5$	X			97
48	Strontium titanate (IV)	$SrTiO_3$		X		102
49	Barium titanate (IV)	$BaTiO_3$	X	X		103
50	Strontium zirconate (IV)	$SrZrO_3$	X			115
51	Lead zirconate (IV)	$PbZrO_3$	X	X		117
52	Zinc zirconate (IV)	$ZnZrO_3$	X			119
53	Cerium zirconate (IV)	$Ce(ZrO_3)_2$	X	X		121
54	Aluminum nitride	AlN	X			133
55	Titanium nitride	Ti_3N_4	X			135
56	Vanadium nitride	VN	X			136
57	Molybdenum nitride	Mo_2N	X			140
58	Sodium azide	NaN_3	X	X	(X)	143
59	Potassium nitrite	$KNO_2 \cdot xH_2O$	X	X		—
60	Lead nitrite	$Pb(NO_2)_2 \cdot xH_2O$		X		151
61	Sodium hexanitrocobaltate (III)	$Na_3Co(NO_2)_6$		X		153
62	Ammonium nitrate	NH_4NO_3	X	X		154
63	Sodium nitrate	$NaNO_3$		X	(X)	155
64	Potassium nitrate	KNO_3	X	X	(X)	156
65	Cesium nitrate	$CsNO_3$	X	X		158
66	Calcium nitrate	$Ca(NO_3) \cdot 4H_2O$		X		159
67	Strontium nitrate	$Sr(NO_3)_2$	X	X		160
68	Aluminum nitrate	$Al(NO_3)_3 \cdot 9H_2O$	X	X		162
69	Thallium nitrate	$Tl(NO_3)_3$	X	X		165
70	Lead nitrate	$Pb(NO_3)_2$	X	X		166
71	Bismuth nitrate	$Bi(NO_3)_3 \cdot 5H_2O$	X	X		168
72	Chromium nitrate	$Cr(NO_3) \cdot 9H_2O$	X			170
73	Iron (III) nitrate	$Fe(NO_3)_3 \cdot xH_2O$	X			171
74	Zinc nitrate	$Zn(NO_3)_2 \cdot xH_2O$	X	X		174
75	Zirconium nitrate	$Zr(NO_3)_4 \cdot 5H_2O$	X			176
76	Silver nitrate	$AgNO_3$	X	X		177

Spectrum Number	Compound	Formula	Spectra IR	R	Vol. 4 IR
77	Lanthanum nitrate	$La(NO_3)_3 \cdot 6H_2O$	X	X	
78	Cerium nitrate	$Ce(NO_3)_3 \cdot 6H_2O$	X	X	181
79	Bismuth subnitrate	$BiONO_3 \cdot H_2O$	X	X	192
80	Tellurium nitrate (basic)	$4TeO_2 \cdot N_2O_5 \cdot 1\ 1/2H_2O$	X	X	193
81	Uranyl nitrate	$UO_2(NO_3)_2 \cdot 6H_2O$	X	X	195
82	Sodium cobaltic nitrate	$NaCo(NO_3)_4$	X		—
83	Neodymium ammonium nitrate	$NdNH_4(NO_3)_4$	X	X	—
84	Sodium hypophosphite	$NaH_2PO_2 \cdot H_2O$	X	X	206
85	Potassium hypophosphite	$KH_2PO_2 \cdot xH_2O$	X	X (X)	207
86	Calcium hypophosphite	$Ca(H_2PO_2)_2$	X	X	208
87	Magnesium hypophosphite	$Mg(H_2PO_2)_2 \cdot xH_2O$	X	X	—
88	Barium hypophosphite	$Ba(H_2PO_2)_2$	X	X	—
89	Iron (III) hypophosphite	$Fe(H_2PO_2)_3$		X	210
89a	Sodium orthophosphite	$Na_2HPO_3 \cdot 5H_2O$		(X)	211
89b	Barium orthophosphite	$BaHPO_3$		(X)	212
90	Sodium metaphosphate	$(NaPO_3)_x \cdot xH_2O$	X	X	214
91	Potassium metaphosphate	$(KPO_3)_x \cdot xH_2O$		X	215
92	Beryllium metaphosphate	$[Be(PO_3)_2]_x \cdot xH_2O$	X	X	216
93	Calcium metaphosphate	$[Ca(PO_3)_2]_x \cdot xH_2O$	X	X	218
94	Strontium metaphosphate	$[Sr(PO_3)_2]_x \cdot xH_2O$	X	X	219
95	Lead metaphosphate	$[Pb(PO_3)_2]_x \cdot xH_2O$	X	X	222
96	Sodium orthophosphate (monobasic)	$NaH_2PO_4 \cdot xH_2O$	X	X	—
96a	Potassium orthophosphate (monobasic)	KH_2PO_4		(X)	225
97	Lead orthophosphate (monobasic)	$Pb(H_2PO_4)_2$		X	—
98	Sodium ammonium orthophosphate (dibasic)	$NaNH_4HPO_4$	X	X	—
99	Barium orthophosphate (dibasic)	$BaHPO_4 \cdot xH_2O$	X		231
100	Lithium orthophosphate	$Li_3PO_4 \cdot 1/2H_2O$		X	234
101	Sodium orthophosphate	$Na_3PO_4 \cdot H_2O$	X		235
102	Magnesium orthophosphate	$Mg_3(PO_4)_2 \cdot 8H_2O$	X		236
103	Barium orthophosphate	$Ba_3(PO_4)_2$		X	—
104	Boron orthophosphate (tetragonal)	BPO_4	X		239
105	Tin (II) orthophosphate	$Sn_3(PO_4)_2$ wet	X		241
106	Lead orthophosphate	$Pb_3(PO_4)_2$ wet	X	X	242b
107	Bismuth orthophosphate	$BiPO_4$	X	X	244
108	Chromium (III) orthophosphate	$CrPO_4 \cdot 6H_2O$	X		245
109	Iron (II) orthophosphate	$Fe_3(PO_4)_2 \cdot 8H_2O$	X		246
110	Iron (III) orthophosphate	$FePO_4 \cdot 2H_2O$	X		247
111	Nickel orthophosphate	$Ni_3(PO_4)_2 \cdot 8H_2O$	X		248
112	Copper (II) orthophosphate	$Cu_3(PO_4) \cdot 3H_2O$	X		249
113	Zinc orthophosphate	$Zn_3(PO_4)_2 \cdot 4H_2O$		X	250
114	Cadmium orthophosphate	$Cd_3(PO_4) \cdot xH_2O$		X	252
115	Magnesium ammonium orthophosphate	$NH_4MgPO_4 \cdot H_2O$		X	254
116	Ammonium cobalt orthophosphate	$NH_4CoPO_4 \cdot xH_2O$	X		256
117	Dilithium sodium orthophosphate	$Li_2NaPO_4 \cdot xH_2O$	X		257

Spectrum Number	Compound	Formula	Spectra IR	R		Vol. 4 IR
159	Cadmium hexahydrostannate (IV)	$CdSn(OH)_6$	X			382
160	Arsenic disulfide	As_2S_2	X			395
161	Antimony trisulfide	Sb_2S_3	X			397
162	Bismuth trisulfide	Bi_2S_3	X			398
163	Tellrurium sulfide	TeS_2	X			399
164	Titanium sesquisulfide	Ti_2S_3	X			400
165	Nickel monosulfide	NiS	X			401
166	Silver sulfide	Ag_2S	X			406
167	Tantalum disulfide	TaS_2	X			409
168	Ammonium imidodisulfate	$(NH_4)_2S_2NHO_6$	X	X		413
169	Sodium hydrogen sulfate	$NaHSO_4 \cdot H_2O$		X	(X)	415
170	Potassium hydrogen sulfate	$KHSO_4$		X		416
171	Potassium thiosulfate	$K_2S_2O_3 \cdot 1/3H_2O$	X	X	(X)	418
172	Barium thiosulfate	$BaS_2O_3 \cdot H_2O$	X	X		420
173	Lead thiosulfate	$PbS_2O_3 \cdot xH_2O$ or wet	X			421
174	Sodium pyrosulfite	$Na_2S_2O_5$	X	X		422
175	Potassium pyrosulfite	$K_2S_2O_5$		X		423
176	Ammonium sulfite	$(NH_4)_2SO_3$		X		—
177	Sodium sulfite	Na_2SO_3		X		424
178	Potassium sulfite	K_2SO_3		X		—
179	Magnesium sulfite	$MgSO_3 \cdot xH_2O$	X	X		425
180	Strontium sulfite	$SrSO_3$		X		426
181	Barium sulfite	$BaSO_3 \cdot xH_2O$ or wet	X	X		427
182	Sodium dithionate	$Na_2S_2O_6$		X		—
183	Potassium dithionate	$K_2S_2O_6$		X		429
184	Silver pyrosulfite	$Ag_2S_2O_7 \cdot xH_2O$	X			430
185	Lithium sulfate	$Li_2SO_4 \cdot H_2O$	X	X		432
186	Sodium sulfate	Na_2SO_4	X	X	(X)	433
187	Potassium sulfate	K_2SO_4	X	X		434
188	Beryllium sulfate	$BeSO_4 \cdot 4H_2O$	X	X		437
189	Magnesium sulfate	$MgSO_4 \cdot 7H_2O$	X	X		439
190	Calcium sulfate	$CaSO_4 \cdot 2H_2O$	X	X		441
191	Strontium sulfate	$SrSO_4$	X	X		442
192	Barium sulfate	$BaSO_4$	X	X		443
193	Aluminum sulfate	$Al_2(SO_4)_3 \cdot 18H_2O$	X	X		444
194	Thallium sulfate	$Tl(SO_4)_3$	X	X		447
195	Lead sulfate tribasic	$3PbO \cdot PbSO_4 \cdot xH_2O$		X		448
196	Antimony sulfate	$Sb_2(SO_4)_3 \cdot xH_2O$	X	X		449
197	Bismuth sulfate	$Bi_2(SO_4)_3 \cdot xH_2O$	X	X		450
198	Vanadyl sulfate	$VSO_4 \cdot xH_2O$	X	X		
199	Iron (III) sulfate	$Fe_2(SO_4)_3 \cdot 9H_2O$		X		454
200	Cobalt (II) sulfate	$CoSO_4 \cdot 7H_2O$	X	X		455
201	Copper (II) sulfate	$CuSO_4 \cdot 5H_2O$	X			457
202	Zinc sulfate	$ZnSO_4 \cdot 6H_2O$	X	X		458
203	Zirconium sulfate	$Zr(SO_4)_2 \cdot 4H_2O$	X	X		461
204	Silver sulfate	Ag_2SO_4	X	X		462
205	Cadmium sulfate	$CdSO_4 \cdot 7H_2O$	X	X		463
206	Cerium (III) sulfate	$Ce_2(SO_4)_3 \cdot 4H_2O$	X	X		467
207	Cerium (IV) sulfate	$Ce(SO_4)_2 \cdot 4H_2O$	X	X		468

Spectrum Number	Compound	Formula	Spectra IR	R		Vol. 4 IR
208	Thorium sulfate	$Th(SO_4)_2 \cdot xH_2O$	X	X		478
209	Ammonium manganese sulfate	$(NH_4)_2MnSO_4 \cdot xH_2O$	X	X		484
210	Ammonium iron (III) sulfate	$(NH_4)Fe(SO_4)_2 \cdot 3H_2O$	X	X		486
211	Ammonium iron (III) sulfate	$NH_4Fe(SO_4)_2 \cdot xH_2O$		X		487
212	Ammonium cobalt sulfate	$(NH_4)_2Co(SO_4)_2 \cdot 6H_2O$	X			488
213	Potassium chromium sulfate	$KCr(SO_4)_2 \cdot 12H_2O$		X		493
214	Aluminum sodium sulfate	$NaAl(SO_4)_2 \cdot xH_2O$	X	X		—
215	Cesium aluminum sulfate	$CsAl(SO_4)_2 \cdot 12H_2O$	X	X		499
216	Ammonium cadmium sulfate	$(NH_4)Cd(SO_4)_2 \cdot 6H_2O$		X		—
216a	Sodium peroxydisulfate	$Na_2S_2O_8$	X	X	(X)	502
217	Potassium peroxydisulfate	$K_2S_2O_8$	X	X		503
218	Gallium monselenide	GaSe	X			506
219	Tin (II) selenide	SnSe	X			507
220	Lead selenide	PbSe	X			508
221	Titatium diselenide	$TiSe_2$	X			509
221a	Sodium selenite	Na_2SeO_3			(X)	517
222	Zinc selenite	$ZnSeO_3$	X			520
223	Copper selenite	$Cu(OH)SeO_3H \cdot H_2O$	X			521
224	Ammonium selenate	$(NH_4)_2SeO_4$		X		522
224a	Sodium selenate	Na_2SeO_4			(X)	523
225	Calcium selenate	$CaSeO_4 \cdot 2H_2O$	X	X		526
226	Copper (II) selenate	$CuSeO_4 \cdot 5H_2O$	X			529
227	Zinc selenate	$ZnSeO_4$		X		—
228	Silver selenate	$Ag_2SeO_4 \cdot xH_2O$ or wet	X	X		530
229	Potassium alumino selenate	$KAl(SeO_4)_2 \cdot 8H_2O$	X	X		—
230	Tin (II) telluride	SnTe	X			531
230a	Zinc telluride	ZnTe	X			536
231	Ammonium dichromate	$(NH_4)_2Cr_2O_7$	X	X		542
232	Lithium dichromate	$Li_2Cr_2O_7 \cdot 2H_2O$	X		(X)	543
233	Sodium dichromate	$Na_2Cr_2O_7 \cdot 2H_2O$	X	X		544
234	Potassium dichromate	$K_2Cr_2O_7$		X		545
235	Calcium dichromate	$CaCr_2O_7 \cdot xH_2O$	X	X		547
236	Silver dichromate	$Ag_2Cr_2O_7 \cdot xH_2O$ or wet	X			549
237	Ammonium chromate	$(NH_4)_2CrO_4 \cdot xH_2O$	X	X		550
238	Lithium chromate	$Li_2CrO_4 \cdot xH_2O$	X	X		551
239	Sodium chromate	$Na_2CrO_4 \cdot xH_2O$		X		552
240	Potassium chromate	K_2CrO_4		X	(X)	553
241	Cesium chromate	Cs_2CrO_4	X	X		554
242	Magnesium chromate	$MgCrO_4$		X		555
243	Calcium chromate	$CaCrO_4$		X		556
244	Aluminum chromate	$Al_2(CrO_4)_3 \cdot xH_2O$	X	X		557
245	Lead chromate	$PbCrO_4$		X		558
246	Cadmium chromate (carbonate imp)	$CdCrO_4$	X			559
247	Potassium zinc chromate	$K_2CrO_4 \cdot 3ZnCrO_4 \cdot Zn(OH)_2$	X	X		561
248	Ammonium molybdate (VI)	$(NH_4)_2MoO_4$	X	X		—
249	Sodium molybdate (VI)	$Na_2MoO_4 \cdot 2H_2O$	X	X	(X)	563
250	Lead molybdate (VI)	$PbMoO_4 \cdot xH_2O$ or wet	X	X		568
251	Nickel molybdate (VI)	$NiMoO_4 \cdot H_2O$ or wet	X			570

Spectrum Number	Compound	Formula	Spectra IR	Spectra R		Vol. 4 IR
252	Copper (II) molybdate (VI)	$CuMoO_4 \cdot xH_2O$	X			571
253	Ammonium phosphomolybdate	$(NH_4)_3PMo_{12}O_{40}$	X	X		577
254	Sodium phosphomolybdate	$Na_3PMo_{12}O_{40}$		X		—
255	Ammonium tungstate	$(NH_4)_2WO_4$	X	X		—
256	Sodium tungstate	$Na_2WO_4 \cdot 2H_2O$	X	X	(X)	579
257	Calcium tungstate	$CaWO_4$	X	X		582
258	Zinc tungstate	$ZnWO_4 \cdot xH_2O$	X	X		587
259	Zirconium tungstate	$Zr(WO_4)_2 \cdot xH_2O$ or wet	X			588
260	Silver tungstate	Ag_2WO_4	X			589
261	Sodium paratungstate	$Na_6W_7O_{24} \cdot 16H_2O$	X	X	(X)	591
262	Potassium tungstate	$K_6W_7O_{24} \cdot xH_2O$	X			—
263	Barium borotungstate	$Ba_3(BW_{12}O_{40})_2 \cdot xH_2O$	X	X		595
264	Ammonium phosphotungstate	$(NH_4)_3PW_{12}O_{40} \cdot 4H_2O$	X	X		596
265	Sodium phosphotungstate	$Na_3PW_{12}O_{40} \cdot xH_2O$		X		597
266	Lead difluoride	PbF_2	X			608
267	Titanium tetrafluoride	$TiF_4 \cdot xH_2O$	X			614
268	Vanadium trifluoride	$VF_3 \cdot 3H_2O$	X			615
269	Cobalt (II) fluoride	$CoF_2 \cdot xH_2O$	X			622
270	Nickel fluoride	$NiF_2 \cdot xH_2O$	X			624
271	Copper (II) fluoride	$CuF_2 \cdot 2H_2O$	X			625
272	Zirconium fluoride	$ZrF_4 \cdot xH_2O$	X			628
273	Lanthanum fluoride	LaF_3	X			631
274	Samarium fluoride	SmF_3	X			634
275	Gadolimium fluoride	GdF_3	X			635
276	Dysprosium fluoride	DyF_3	X			636
277	Holmium fluoride	HoF_3		X		637
278	Erbium fluoride	ErF_3	X			638
279	Thorium fluoride	$ThF_4 \cdot 4H_2O$	X			640
280	Ammonium tetrafluoroborate	NH_4BF_4	X	X		642
281	Sodium tetrafluoroborate	$NaBF_4 \cdot xH_2O$	X		(X)	644
282	Ammonium tetrafluoroaluminate	NH_4AlF_4	X			649
283	Ammonium hexafluorogallate	$(NH_4)_3GaF_6$	X			652
284	Ammonium hexafluorosilicate	$(NH_4)_2SiF_6$		X		653
285	Sodium hexafluorosilicate	Na_2SiF_6		X		655
286	Ammonium hexafluorogermanate	$(NH_4)_2GeF_6$	X			665
287	Barium hexafluorogermanate	$BaGeF_6$	X			667
288	Ammonium hexafluorophosphate	NH_4PF_6		X		681
289	Potassium hexafluorophosphate	$KPF_6 \cdot xH_2O$ or wet	X	X	(X)	682
290	Potassium hexafluoroarsenate	$KAsF_6$	X	X	(X)	685
291	Potassium hexafluorotitanate (IV)	K_2TiF_6	X			692
292	Potassium tetrafluorozincate	$K_2ZnF_4 \cdot xH_2O$	X			703
293	Sodium pentafluorozirconate	$NaZrF_5 \cdot xH_2O$	X	X		704
294	Potassium pentafluorozirconate	$KZrF_5 \cdot xH_2O$	X			705
295	Ammonium hexafluorozirconate	$(NH_4)_2ZrF_6$	X	X		706
296	Sodium hexafluorozirconate	Na_2ZrF_6	X	X		707
297	Potassium hexafluorozirconate	$K_2ZrF_6 \cdot xH_2O$	X			708
298	Indium hexafluorozirconate	$In(ZrF_6)_3 \cdot xH_2O$	X			709
299	Potassium heptafluoroniobate (V)	K_2NbF_7		X		711
300	Potassium heptafluorotantalate	K_3TaF_7	X	X		713

Spectrum Number	Compound	Formula	Spectra IR	R		Vol. 4 IR
301	Sodium pentafluorouranate	$NaUF_5 \cdot H_2O$	X			714
302	Potassium oxopentafluoroniobate	$K_2NbOF_5 \cdot xH_2O$	X	X		—
303	Potassium oxohexafluorozirconate	K_3ZrOF_6		X		—
304	Ammonium chloride	NH_4Cl	X	X		715
304a	Sodium chlorite	$NaClO_2$			(X)	779
305	Sodium chlorate	$NaClO_3$		X	(X)	780
306	Strontium chlorate	$Sr(ClO_3)_2 \cdot xH_2O$	X	X		782
307	Barium chlorate	$Ba(ClO_3)_2 \cdot H_2O$	X	X		783
308	Ammonium perchlorate	NH_4ClO_4		X		784
309	Sodium perchlorate	$NaClO_4 \cdot H_2O$	X	X	(X)	786
310	Rubidium perchlorate	$RbClO_4$	X			787
311	Cesium perchlorate	$CsClO_4$	X	X		788
312	Magnesium perchlorate	$Mg(ClO_4)_2 \cdot 6H_2O$		X		789
313	Zinc perchlorate	$Zn(ClO_4)_2 \cdot 6H_2O$		X		793
314	Ammonium bromide	NH_4Br	X	X		795
315	Barium bromide	$BaBr_2 \cdot 2H_2O$	X	X		801
316	Antimony bromide	$SbBr_3$	X			806
317	Bismuth bromide	$BiBr_3 \cdot xH_2O$	X			807
318	Tellurium bromide	$TeBr_4 \cdot xH_2O$	X			808
319	Iron (II) bromide	$FeBr_2 \cdot xH_2O$	X			809
320	Cadmium bromide	$CdBr_2$	X			812
321	Mercury (I) bromide	Hg_2Br_2		X		814
322	Sodium bromate	$NaBrO_3$		X	(X)	819
323	Potassium bromate	$KBrO_3$	X	X		820
324	Rubidium bromate	$RbBrO_3$	X	X		821
325	Cesium bromate	$CsBrO_3$		X		822
326	Magnesium bromate	$Mg(BrO_3)_2 \cdot 6H_2O$	X	X		823
327	Barium bromate	$Ba(BrO_3)_2 \cdot H_2O$		X		824
328	Aluminum bromate	$Al(BrO_3)_3 \cdot 9H_2O$	X	X		825
329	Lead bromate	$Pb(BrO_3)_2 \cdot H_2O$		X		826
330	Cadmium bromate	$Cd(BrO_3)_2 \cdot xH_2O$		X		828
331	Potassium iodide	KI	X			831
332	Rubidium iodide	RbI	X			832
333	Barium iodide	$BaI_2 \cdot 2H_2O$	X			834
334	Thallium iodide	TlI	X			835
335	Arsenic iodide	AsI_3	X			839
336	Antimony iodide	SbI_3	X			840
337	Bismuth iodide	$BiI_3 \cdot xH_2O$ or wet	X			841
338	Zirconium iodide	$ZrI_4 \cdot xH_2O$	X			844
339	Palladium iodide	PdI_2	X			846
340	Silver iodide	AgI_2	X			847
341	Mercury (II) iodide	HgI	X			849
342	Ytterbium iodide	$YbI_3 \cdot xH_2O$	X			850
343	Potassium bismuth iodide	$K_4BiI_7 \cdot xH_2O$	X			851
344	Potassium tetraiodomercurate (II)	$K_2HgI_4 \cdot xH_2O$	X			—
345	Potassium iodocadmate	$K_2CdI_4 \cdot xH_2O$	X			853
346	Ammonium iodate	NH_4IO_3	X	X		854
347	Sodium iodate	$NaIO_3$	X	X		856
348	Rubidium iodate	$RbIO_3$	X	X		858

Spectrum Number	Compound	Formula	Spectra IR	R	Vol. 4 IR
349	Cesium iodate	$CsIO_3 \cdot xH_2O$	X	X	859
350	Calcium iodate	$Ca(IO_3)_2 \cdot 6H_2O$	X	X	860
351	Strontium iodate	$Sr(IO_3)_2 \cdot xH_2O$	X	X	861
352	Barium iodate	$Ba(IO_3)_2 \cdot H_2O$		X	862
353	Lead iodate	$Pb(IO_3)_2$	X	X	863
354	Chromium (III) iodate	$Cr(IO_3)_3 \cdot xH_2O$	X	X	864
355	Nickel iodate	$Ni(IO_3)_2 \cdot xH_2O$	X		865
356	Silver iodate	$AgIO_3$	X	X	866
357	Cerium iodate	$Ce(IO_3)_4 \cdot xH_2O$	X		867
358	Cobalt iodate	$Co(IO_3)_2 \cdot H_2O$	X		—
359	Zinc iodate	$Zn(IO_3)_x \cdot xH_2O$	X	X	—
360	Sodium periodate	$NaIO_4$		X	868
361	Potassium periodate	KIO_4		X (X)	869
362	Potassium permanganate	$KMnO_4$	X	(X)	874

(X) water solution.
Vols. 2–3 = Nyquist, Putzig, and Leugers.
Vol. 4 = Nyquist and Kagel, "Infrared Spectra of Inorganic Compounds: 3800-45cm⁻¹".

B. NUMERICAL INDEX OF RAMAN (R) AND INFRARED (IR) SPECTRA OF NONIONIC INORGANIC COMPOUNDS AND HYDRAZINE SALTS

Spectrum Number	Compound	Formula	Spectra IR	R	Vol. 4 IR
	Boron				
363	Diborane	B_2H_6	XX		
364	Decaborane	$B_{10}H_{14}$	XX		
365	Hydrazine diborane	$B_2H_{10}N_2$	X		
366	Decaborane monohydrazine	$B_{10}H_{16}N_2$	X		
367	Boron trifluoride	BF_3	X		
368	Boron tribromide	BBr_3	X		
	Carbon				
369	Carbon monoxide	CO	X		
370	Carbon dioxide	CO_2	X		
371	Carbonyl sulfide	COS	X		
372	Carbon disulfide	CS_2	X	X	
373	Hydrogen cyanide	CHN	X		
374	Cyanogen	C_2N_2	X		
375	Cyanogen chloride	$CClN$	X		
376	Cyanogen bromide	$CBrN$	XX		
377	Cyanamide	CH_2N_2	X		
378	Cyanoquanidine	$C_2H_4N_4$	X		
379	Carbonyl fluoride	CF_2O	X		
380	Carbonyl chloride	CCl_2O	X		
	Nitrogen				

Spectrum Number	Compound	Formula	Spectra IR	R	Vol. 4 IR
381	Ammonia	NH$_3$	X		
382	Ammonia-d$_3$	ND$_3$	X		
383	Nitrous oxide	N$_2$O	X		
384	Nitric oxide	NO plus NO$_2$	X		
385	Nitrogen trifluoride	NF$_3$	X		
386	Nitrogen trichloride	NCl$_3$	X		
387	Hydrazine tetrafluoride	N$_2$F$_4$	X		
388	Thiazyl trifluoride	NF$_3$S	X		
389	Hydrazine dichloride	N$_2$H$_6$Cl$_2$	X		
	Phosphorus				
390	Phosphine	PH$_3$	XX		
391	Phosphorus trifluoride	PF$_3$	XX		
392	Phosphorus trichloride	PCl$_3$	XXXX		
393	Phosphorus tribromide	PBr$_3$	XX		
394	Phosphorus triiodide	PI$_3$	XX		
395	Phosphorus oxyfluoride	PF$_3$O	XX		
396	Phosphorus oxychloride	PCl$_3$O	XX		
397	Phosphorus oxybromide	PBr$_3$O	XX		
398	Thiophosphoryl dichloride fluoride	PCl$_2$FS	X	(R)	
399	Thiophosphoryl chloride	PCl$_3$S	XXX		
400	Thiophosphoryl bromide	PBr$_3$S	XX		
401	Phosphorus pentachloride	PCl$_5$	X	(R)	
402	Phosphorus pentabromide	PBr$_5$	XX		
403	Trichlorophosphazosulfuryl chloride	PCl$_4$NO$_2$S			
404	Phosphoric acid	PH$_3$O$_4$	X		
405	Phosphorus thioamide	PH$_6$N$_3$S	X		
406	Phosphorus pentasulfide	P$_2$S$_5$	X		
407	Chloromethylphosphonic dichloride	PCH$_2$Cl$_3$O	XX		
	Arsenic				
408	Arsine	AsH$_3$	X		
	Oxygen				
409	Ozone	O$_3$	X		
410	Water	OH$_2$	XX		
411	Deuteruim oxide	OD$_2$	X		
	Sulfur				
412	Sulfur	S$_8$	XX	X	811a
413	Hydrogen sulfide	SH$_2$	X		
414	Sulfur dioxide	SO$_2$	X		
415	Sulfur monochloride	S$_2$Cl$_2$	X		
416	Sulfur dichloride	SCl$_2$	X		
417	Sulfur monobromide	S$_2$Br$_2$	XX		
418	Thionyl fluoride	SF$_2$O	XX		
419	Thionyl chloride	SCl$_2$O	XXX	X	
420	Sulfuryl chloride	SCl$_2$O$_2$	XXX		
421	Pyrosulfuryl chloride	S$_2$Cl$_2$O$_5$	X		

Spectrum Number	Compound	Formula	Spectra IR	R	Vol. 4 IR
422	β-Sulfanuryl chloride	$S_3Cl_3N_3O_3$	X		
423	Sulfonamide	$SH_4N_2O_2$	X		
424	Sulfuric acid	SH_2O_4	X		
425	Sulfur hexafluoride	SF_6	X		
426	Decafluorodisulfide	S_2F_{10}	X		
427	Methyl chlorosulfinate	CH_3ClO_2S	X		
428	Potassium methyl sulfate	CH_3OSO_3K		X	

Halogen

429	Chlorine	Cl_2		(R)	
430	Bromine	Br_2		(R)	
431	Iodine	I_2		(R)	
432	Hydrogen chloride	ClH	X		
433	Deuterium chloride	ClD	X		
434	Hydrogen bromide	BrH	X		
435	Deuterium bromide	BrD	X		

(R) Recorded using a Spex Ramalog instrument equipped with a Spectra-Physics Model 125 He-Ne laser (70 mW at 6328A).

Vols. 2–3 = Nyquist, Putzig, and Leugers.

Vol. 4 = Nyquist and Kagel, "Infrared Spectra of Inorganic Compounds: 3800-45cm^{-1}".

C. NUMERICAL INDEX OF MISCELLANEOUS MINERALS

Spectrum Number	Compound	Formula	Spectra IR	R	Vol. 4 IR
436	Aragonite	$CaCO_3$	X	X	—
437	Asbestos (amphibole)		X		—
438	Asbestos (crocidolite)		X		—
439	Asbestos (serpentine chrysolite)		X		—
440	Gearksutite	$CaAlF(OH)$	X	X	—
441	Hydromagnesite	$3Mg \cdot Mg(OH)_2 \cdot 3H_2O$	X	X	—
442	Itacolumite		X	X	—
443	Kalinite	$AlK(SO_4)_2 \cdot 12H_2O$	X	X	—
444	Meyerhofferite	$2Ca \cdot 3B_2O_3 \cdot 7H_2O$	X	X	—
445	Quartz	SiO_2	X	X	892
446	Realgar	AsS	X	X	—
447	Stilbite	$H_4(Na_2,Ca)Al_2Si_6O_{18} \cdot 4H_2O$	X	X	—
448	Thaumasite	$CaSiO_3 \cdot CaCO_3 \cdot CaSO_4 \cdot 15H_2OX$	X	X	—
449	Vermiculite		X		—

D. NUMERICAL INDEX OF ORGANIC SALTS (OS) CARBOXYLIC ACID SALTS, PHENATES, AND BENZENESULFONATES

Spectrum Number OS	Compound	Formula	Spectra R	IR
1	Sodium formate	CHO_2Na	X	X
2	Magnesium formate	$(CHO_2)_2Mg \cdot xH_2O$	X	X
3	Calcium formate	$(CHO_2)_2Ca$	X	X
4	Strontium formate	$(CHO_2)_2Sr$	X	X
5	Barium formate	$(CHO_2)_2Ba$	X	X
6	Lead formate	$(CHO_2)_2Pb$	X	X
7	Zinc formate	$(CHO_2)_2Zn \cdot xH_2O$	X	X
8	Sodium acetate	$(CH_3CO_2)_2Na \cdot xH_2O$	X	X
9	Calcium acetate	$(CH_3CO_2)_2Ca \cdot xH_2O$	X	X
10	Strontium acetate	$(CH_3CO_2)_2Sr \cdot xH_2O$	X	X
11	Barium acetate	$(CH_3CO_2)_2Ba \cdot xH_2O$	X	X
12	Thallous acetate	$(CH_3CO_2)Tl \cdot xH_2O$	X	X
13	Guanidinium acetate	$(CH_3CO_2)(H_2N)_2C=NH_2$	X	X
14	Sodium butyrate	$(n\text{-}C_3H_7CO_2)Na$	X	X
15	Calcium butyrate	$(n\text{-}C_3H_7CO_2)_2Ca$	X	X
16	Sodium valerate	$(n\text{-}C_4H_9CO_2)Na \cdot xH_2O$	X	X
17	Potassium valerate	$(n\text{-}C_4H_9CO_2)K$	X	X
18	Sodium stearate	$(n\text{-}C_{17}H_{35}CO_2)Na$	X	X
19	Zinc stearate	$(n\text{-}C_{17}H_{35}CO_2)_2Zn$	X	X
20	Sodium cyanoacetate	$(NCCH_2CO_2)Na$	X	X
21	Lithium oxalate	$(C_2O_4)Li_2$	X	X
22	Sodium oxalate	$(C_2O_4)Na_2$	X	X
23	Potassium oxalate hydrate	$(C_2O_4)K_2 \cdot xH_2O$	X	
24	Calcium oxalate	$(C_2O_4)Ca$	X	X
25	Strontium oxalate	$(C_2O_4)Sr \cdot xH_2O$	X	X
26	Barium oxalate	$(C_2O_4)Ba \cdot xH_2O$	X	X
27	Lead oxalate	$(C_2O_4)Pb$	X	X
28	Manganese oxalate	$(C_2O_4)Mn \cdot xH_2O$	X	X
29	Stannous oxalate	$(C_2O_4)Sn$	X	
30	Bismuth oxalate	$(C_2O_4)Bi \cdot xH_2O$	X	X
31	Cadmium oxalate	$(C_2O_4)Cd$	X	X
32	Thallous malonate	$(CH_2(CO_2)_2)Tl_2 \cdot xH_2O$	X	X
33	Sodium succinate	$(CH_2CO_2)_2Na_2 \cdot xH_2O$	X	X
34	Potassium tartrate	$(CHOHCO_2)_2K_2$	X	
35	Strontium tartrate	$(CHOHCO_2)_2Sr$	X	X
36	Barium tartrate	$(CHOHCO_2)_2Ba$	X	X
37	Lead tartrate	$(CHOHCO_2)_2 Pb \cdot xH_2O$	X	X
38	Lithium citrate	$((O_2CCH_2)C(OH)(CO_2))Li3$	X	X
39	Sodium citrate	$((O_2CCH_2)_2C(OH)(CO_2))Na_3$	X	X
40	Calcium citrate	$((O_2CCH_2)_2C(OH)(CO_2))_2Ca_3$	X	
41	Bismuth citrate	$((O_2CCH_2)_2C(OH)(CO_2))Bi_3$	X	X
42	Manganese citrate	$((O_2CCH_2)_2C(OH)(CO_2))_2Mn$	X	X
43	Bismuth ammonium citrate	$((O_2CCH_2)C(OH)(CO_2))BeNH_4$	X	X
44	Manganese sodium citrate	$((O_2CCH_2)C(OH)(CO_2))MnNa$	X	X
45	Stannous EDTA	$[CH_2\text{-}N(\text{-}CH_2\text{-}CO_2)]_2Sn_2$	X	
46	Lithium hippurate	$(C_6H_5C(=O)NHCH_2CO_2)Li$	X	X

Spectrum Number OS	Compound	Formula	Spectra R	IR
47	Sodium hippurate	$(C_6H_5C(\!=\!O)NHCH_2CO_2)Na$	X	X
48	Calcium hippurate	$[((C_6H_5C(\!=\!O)NHCH_2CO_2)]Ca$	X	X
49	Lithium benzoate	$(C_6H_5CO_2)Li$	X	X
50	Sodium benzoate	$(C_6H_5CO_2)Na$	X	X
51	Calcium benzoate	$(C_6H_5CO_2)_2Ca$	X	X
52	Bismuth benzoate	$(C_6H_5CO_2)_2Bi$	X	X
53	Manganese benzoate	$(C_6H_5CO_2)_2Mn$	X	X
54	Zinc benzoate	$(C_6H_5CO_2)_2Zn$	X	X
55	Lead benzoate	$(C_6H_5CO_2)_2Pb$	X	X
56	Ammonium salicylate	$(o\text{-}(OH)C_6H_4CO_2)NH_4$	X	X
57	Lithium salicylate	$(o\text{-}(OH)C_6H_4CO_2)Li$	X	X
58	Sodium salicylate	$(o\text{-}(OH)C_6H_4CO_2)Na$	X	X
59	Calcium salicylate	$(o\text{-}(OH)C_6H_4CO_2)_2Ca$	X	X
60	Bismuth salicylate	$(o\text{-}(OH)C_6H_4CO_2)_2Bi$	X	X
61	Ferric salicylate	$(o\text{-}(OH)C_6H_4CO_2)_3Fe$	X	
62	Zinc salicylate	$(o\text{-}(OH)C_6H_4CO_2)_2Zn$	X	X
63	Cadmium salicylate	$(o\text{-}(OH)C_6H_4CO_2)_2Cd$	X	X
64	Potassium phthalate	$[o\text{-}C_6H_4(CO_2)_2]K_2$	X	X
65	Sodium cinnamate	$(C_6H_5CH\!=\!CHCO_2)Na$	X	X
66	Sodium 4-nitrophenate	$(4\text{-}NO_2C_6H_4O)Na$	X	
67	Sodium 2,4-dinitrophenate	$(2,4\text{-}(NO_2)_2C_6H_3O)Na$	X	
68	Sodium benzenesulfonate	$(C_6H_5SO_3)Na$	X	X
69	Sodium 4-hydroxybenzenesulfonate	$(4\text{-}(OH)C_6H_4SO_3)Na$	X	

E. NUMERICAL INDEX OF INORGANIC MATERIALS USED FOR INFRARED WINDOWS

Spectrum Number	Window Material	Spectra IR
W1	Pyrex	X
W2	Quartz	X
W3	Silicon wafer	X
W4	Silicon carbide	X
W5	Lithium fluoride	X
W6	Sodium chloride	X
W7	Potassium bromide	X
W8	Silver chloride	X
W9	Zinc selinide	X
W10	Cadmium telluride	X
W11	Barium fluoride sealed cell (0.026 mm pathlength)	X
W12	Sodium chloride sealed cell (0.108 mm pathlength)	X
W13	Calcium fluoride sealed cell (0.099 mm pathlength)	X
W14	Potassium chloride sealed cell (0.200 mm pathlength)	X

A. ALPHABETICAL INDEX OF SPECTRA FOR IONIC INORGANIC COMPOUNDS AND MINERALS

Compound	Vols. 2–3 Spectrum Number			Vol. 4 Spectrum Number
	IR	R	R in water solution	
Alum, potassium basic				500
Aluminate				
lithium				24
potassium				25
hexafluoro-				
ammonium				650
potassium				651
tetrafluoro-, ammonium	282			649
tetrahydro-, lithium				23
Antimonate				
lead				304
hexafluoro-				
potassium				687
silver				688
hexahydroxo-				
potassium				384
sodium				383
tetrafluoro-, ammonium				686
Argenate (l)				
cyano-, potassium	13	13		33
Arsenate				
hexafluoro-				
potassium	290	290		685
ortho-				
ammonium (dibasic)	130	130		296
antimony	131	131		298
cobalt	133			300
copper (II)	134			301
iron (III)	132			299
mercury (II)		136		303

Compound	Vols. 2–3 Spectrum Number			Vol. 4 Spectrum Number
	IR	R	R in water solution	
potassium (monobasic)			(129a)	295
sodium (dibasic)				297
			R (aqueous)	
zinc	135	135		302
pyro-, lead		129		294
Arsenide, manganese				285
meta-				
lead				287
sodium			(125a)	286
zinc	126			288
ortho-				
antimony	127			289
copper (I)				291
iron (III)	128			290
mercury (I)				293
silver				292
Azide				
ammonium				142
barium				147
cesium				146
potassium				144
rubidium				145
sodium	58	58	(58)	143
Bicarbonate, *see* Carbonate				
Borate				
per-, sodium				17
tetra-				
lithium	2			18
potassium	4	4		22
sodium	3	3		19, 20, 21
tetrafluoro-				
ammonium	280	280		642
calcium				646
lithium				643
nickel				647
potassium				645
sodium	281		(281)	644
zinc				648
Boric Acid	1	1		16
Boride				
aluminum				1
calcium				3
chromium				6, 7, 8
lanthanum				12
magnesium				2
molybdenum				11
niobium				10

Compound	Vols. 2–3 Spectrum Number			Vol. 4 Spectrum Number
	IR	R	R in water solution	
silicon				4
tantalum				3
tungsten				14, 15
vanadium				5
zirconium				9
Borotungstate, *see* Tungsten				
Bromate				
aluminum	328	328		825
barium		327		824
cadmium		330		828
cesium		325		822
lead		329		826
lithium				818
magnesium	326	326		823
potassium	323	323		820
rubidium	324	324		821
sodium		322	(322)	819
zinc				827
Bromide				
ammonium		314		795
ammonium cadmium				817
antimony (III)	316			806
arsenic				805
barium	315	315		801
bismuth	317			807
cadmium	320			812
cesium				799
holmium				816
indium				802
iron (II)	319			809
lanthanum				813
lead				804
mercury (I),(II)		321		814, 814a
neodymium				815
potassium				797
rubidium				798
silver				811
sodium				796
strontium				800
tellurium	318			808
tin				803
zinc				810
Cadmate				
iodo-, potassium	345			853
Carbonate				
barium		34		63

Compound	Vols. 2–3 Spectrum Number			Vol. 4 Spectrum Number
	IR	R	R in water solution	
bismuth (basic)				70
cadmium	36	36		68
calcium		33		61, 61a
cesium				60
cobalt (basic)				66
copper (II) (basic)				72, 73
guanidinium	37	37		
lead	64	35	35	
lead (basic)				69
lithium	31	31		56
manganese				65
nickel				71
potassium			(32a)	59
silver				67
sodium	32	32	(32)	57, 58
strontium				62
zinc (basic)				74
bi-				
ammonium				53
potassium	30	30		55
sodium	29	29	(29)	54
thio-, barium	38	38		75
Carbide, boron				26
Cerate (III)				
pentanitrato-, ammonium				196
Cerate (IV)				
hexanitrato-				
ammonium				197
magnesium				199
potassium				198
Chlorate				
barium		307		783
potassium				781
sodium		305		780
strontium	306	306		782
per-				
ammonium		308		784
barium				790, 791
cerium	311	311		794
cesium				788
gallium				792
lithium				785
magnesium	312	312		789
rubidium	310			787
sodium	309	309	(309)	786
zinc		313		793

Compound	Vols. 2–3 Spectrum Number			Vol. 4 Spectrum Number
	IR	R	R in water solution	
Chloride				
aluminum				726
ammonium		304		715
ammonium gallium				762
ammonium magnesium				758
barium				725
barium cadmium				776
cadmium				742
cerium				749
calcium				722, 723
cesium				720
chromium (III)				732, 733
cobalt				735
gadolinium				752
hafnium				744
hexammine cobalt				757
holmium				753
indium				727, 728
iron (II)				734
lanthanum				743
lead				730
lead fluoride				610
lithium				716
magnesium				721
mercury (I)				747
mercury (II)				748
mercury amide				755
nickel				736
niobium				739
palladium				740
palladium, diammine (trans)				756
potassium				718
potassium magnesium				759
praseodymium				750
rubidium				719
samarium				751
silver				741
sodium				717
sodium aluminum				761
strontium				724
tantalum				745
thallium				729
thorium				754
tungsten				746
uranyl				394
vanadium				731
yttrium				738

Compound	Vols. 2–3 Spectrum Number			Vol. 4 Spectrum Number
	IR	R	R in water solution	
zinc				737
zirconyl				391
oxy-				
antimony				387
bismuth				388
molybdenum				392
Chlorite, sodium			(304a)	779
Chromate				
aluminum	244	244		557
ammonium	237	237		550
cadmium	246			559
calcium		243		556
cesium	221	221		554
lead		245		558
lithium	238	238	(238)	551
lithium sodium				560
magnesium		242		555
potassium		240	(240)	553
potassium zinc	247	247		561
sodium	239	239		552
di-				
ammonium	231	231		542
calcium	547	235	235	
lithium	232			543
potassium		234		545
rubidium				546
silver	236			549
sodium	233	233		544
zinc				548
hexafluoro-, potassium				696
Chromite				
copper (I)				540
copper (II)				541
Clay, kaolin				95
Cobaltite, lithium				881
Cobaltite (III)				
hexanitro-, sodium		61		153
Cuprate				
cyano-, potassium		14	14	34
tetrachloro-				
ammonium				769
potassium				770
Cyanamide, lead				52
Cyanate				
silver	24	24		45
sodium	23	23		44
thio-				

Compound	Vols. 2–3 Spectrum Number			Vol. 4 Spectrum Number
	IR	R	R in water solution	
cuprous	27	27		49
iron (II)	26			48
lead	25	25		47
mercury (II)				51
potassium	24a		(24a)	46
silver	28	28		50
Cyanide				
cuprous	8	8		28
mercury(II)		12		32
nickel	7	7		27
platinous	11	11		31
potassium		6		—
silver	10			30
sodium		5		—
zinc	9	9		29
ferri-, potassium			(14a)	35
ferro-				
calcium		19		—
ferric	22			42
lead	21	21		41
potassium	17	17		38
potassium calcium	18	18		39
potassium cupric	20	20		40
sodium		16		37
nitroferri-, sodium		15		36
Cyanoargenetate (I), *see* Argentate (I)				
Cyanocuprate, *see* Cuprate				
Cyanoplatinate, *see* Platinate				
Dichromate, *see* Chromate				
Dithionate, *see* Thionate				
Ferrate				
cobalt				878
copper (I)				880
nickel				870
hexafluoro-				
ammonium				701
sodium				702
pentachloro-				
ammonium				767
potassium				768
pentafluoro-, potassium				700
Ferricyanide, *see* Cyanide				
Ferrocyanide, *see* Cyanide				
Fluorodate				
phosphoro-				
barium				283
potassium				282

Compound	Vols. 2–3 Spectrum Number			Vol. 4 Spectrum Number
	IR	R	R in water solution	
sodium				281
phosphorodi-, potassium				284
Fluoride				
aluminium				605
antimony				611
barium				604
bismuth				612
cadmium				630
calcium				602
cerium (III)				633
chromium (III)				617, 618
cobalt (II)	269			622
cobalt (III)				623
copper (II)	271			625
dysprosium	276			636
erbium	278			638
gadolinium	275			635
gallium				606
hafnium				632
holmium		277		637
iron (II)				620
iron (III)				621
lanthanum	273			631
lead (II)	266			608
lead (IV)				609
lithium				598
magnesium				601
manganese				619
nickel	270			624
potassium				600
samarium	274			634
silver				629
sodium				599
strontium				603
thallium				607
thorium	279			640
titanium (III)				613
titanium (IV)	267			614
uranyl				393
vanadium (III)	268			615
vanadium (IV)				616
ytterbium				639
yttrium				627
zinc				626
zirconium	272			628
hydrogen, sodium				641
Fluorosulfonate, *see* Sulfonate				

Compound	Vols. 2–3 Spectrum Number			Vol. 4 Spectrum Number
	IR	R	R in water solution	
Gallate				
hexafluoro-, ammonium	283			652
Germanate				
hexafluoro-				
ammonium	286			665
barium	287			667
sodium				666
Heptafluoroniobate, *see* Niobate				
Heptafluorotantalate, *see* Tantalate				
Heptafluorozirconate, *see* Zirconate				
Hexachloromolybdate, *see* Molybdate				
Hexachloropalladate, *see* Palladate				
Hexachlorostannate, *see* Stannate				
Hexafluoroaluminate, *see* Aluminate				
Hexafluoroantimonate, *see* Antimonate				
Hexafluoroarsenate, *see* Arsenate				
Hexafluorochromate, *see* Chromate				
Hexafluoroferrate, *see* Ferrate				
Hexafluorogallate, *see* Gallate				
Hexafluorogermanate, *see* Germanate				
Hexafluoromanganate, *see* Manganate				
Hexafluorophosphate, *see* Phosphate				
Hexafluorosilicate, *see* Silicate				
Hexafluorostannate, *see* Stannate				
Hexafluorotantalate, *see* Tantalate				
Hexaflorotitanate, *see* Titanate				
Hexafluorozirconate, *see* Zirconate				
Hexahydroxoantimonate, *see* Antimonate				
Hexahydroxostannate, *see* Stannate				
Hexammine cobalt chloride, *see* Chloride				
Hexanitratocerate (IV), *see* Cerate (IV)				
Hexanitrocobaltate, *see* Cobaltate				
Hydrogen fluoride, *see* Fluoride				
Hydroxide				
aluminum (gibbsite)	155			376
barium	154			375
lanthanum				378
lithium	153			372
magnesium				374
nickel	156			377
sodium				373
ammonium, hydrochloride				371
oxy-				
aluminum (boehmite)				385
iron (III)				386
Hyponitrite, *see* Nitrite				
Hypophosphite, *see* Phosphite				

Compound	Vols. 2–3 Spectrum Number			Vol. 4 Spectrum Number
	IR	R	R in water solution	
Imido disulfate, *see* Sulfate				
Iodate				
ammonium	346	346		854
barium		352		862
calcium	350	350		860
cesium	349	349		859
cerium	357			867
chromium (III)	354	354		864
cobalt	358			—
lead	353	353		863
lithium				855
nickel	355			865
rubidium	348	348		858
silver	356	356		866
sodium	347	347		856, 857
strontium	351	351		861
zinc	359	359		—
per-				
potassium		361	(361)	869
sodium		360		868
Iodide				
ammonium				829
antimony	336			840
arsenic	335			839
barium	333			834
bismuth	337			841
cesium				833
copper				843
germanium				836
lead				838
mercury (I)				848
mercury (II)	371			849
nickel				842
niobium				845
palladium	339			846
potassium	331			831
potassium bismuth	343			851
rubidium	332			832
silver	340			847
thallium	334			835
tin (IV)				837
ytterbium	342			850
zirconium	338			844
oxy-, bismuth				390
Iodocadmate, *see* Cadmate				
Kaolin clay, *see* Clay				
Manganate				

Compound	Vols. 2–3 Spectrum Number			Vol. 4 Spectrum Number
	IR	R	R in water solution	
barium				871
hexafluoro-, potassium				697, 698, 699
per-				
barium				876
lithium				872
magnesium				875
potassium	362			874
sodium				873
zinc				877
Manganite, lithium				870
Mercurate				
tetraiodo-, copper	344			852
Metaarsenite, *see* Arsenite				
Metaphosphate, *see* Phosphate				
Metaphosphoric acid, *see* Phosphoric acid				
Metaniobate, *see* Niobate				
Metasilicate, *see* Silicate				
Metavanadate, *see* Vanadate				
Minerals				
albite				885
apatite				886
dolomite				887
hectorite				888
microcline				889
pyrite				890
serpentine				891
wavellite				892
Molybdate (VI)				
ammonium		248		—
barium				567
cadmium				575
calcium				565
cobalt				569
copper	252			571
lead	250	250		568
lithium				562
nickel	251			570
potassium				564
silver				574
sodium	249	249	(249)	563
stontium				566
zinc				572
zirconium				573
hexachloro-, potassium				771
para-, ammonium				576
phospho-, ammonium	253	253		577
phospho-, sodium		254		—

Compound	Vols. 2–3 Spectrum Number			Vol. 4 Spectrum Number
	IR	R	R in water solution	
Niobate				
heptafluoro-, potassium		299		711
meta-, potassium				314
ortho-, potassium				315
oxy-, pentafluoro	302	302		—
Nitrate				
aluminum	68	68		162
ammonium	62	62		154
ammonium neodymium	83	83		—
barium				161
bismuth	71	71		168
cadmium				178
calcium		66		159
cerium	78	78		181
cesium	65	65		158
chromium	72			170
cobalt				172
dysprosium				186
erbium				188
gadolinium				184
gallium				163
holmium				187
indium				164
iron (III)	73			171
lanthanum	77	77		179
lead	70	70		166, 167
mercurous				180
neodymium				182
nickelous				173
potassium		63	(63)	156
rubidium				157
samarium				183
scandium				169
silver	76	76		177
sodium		82		155
sodium cobaltic		63	(63)	—
strontium	67	67		160
tellurium (basic)	80	80		193
terbium				185
thallium	69	69		165
thorium				191
thulium				189
uranyl	81	81		195
ytterbium				190
yttrium				175
zinc	74	74		174
zirconium	75			176

Compound	Vols. 2–3 Spectrum Number			Vol. 4 Spectrum Number
	IR	R	R in water solution	
zirconyl				194
sub-, bismuth	79	79		192
Nitride				
aluminum	54			133
barium				131
boron				132
calcium				130
chromium				137
molybdenum	57			140
niobium				139
silicon				134
tantalum				141
titanium	55			135
vanadium	56			136
zirconium				138
Nitrite				
barium				150
lead		60		151
potassium	59	59		—
silver				152
sodium				149
hypo-, sodium				148
strontium	67	67		160
tellurium (basic)	80	80		193
terbium				185
thallium	69	69		165
thorium				191
thulium				189
uranyl	81	81		195
ytterbium				190
yttrium				175
zinc	74	74		174
zirconium	75			176
zirconyl				194
sub-, bismuth	79	79		192
Nitride				
aluminum	54			133
barium				131
boron				132
calcium				130
chromium				137
molybdenum	57			140
niobium				139
silicon				134
tantalum				141

Compound	Vols. 2–3 Spectrum Number			Vol. 4 Spectrum Number
	IR	R	R in water solution	
titanium	55			135
vanadium	56			136
zirconium				138
Nitrite				
barium				150
lead		60		151
potassium	59	59		—
silver				152
sodium				149
hypo-, sodium				148
Nitroferricyanide, *see* Cyanide				
Orthoarsenate, *see* Arsenate				
Orthoarsenite, *see* Arsenite				
Orthoniobate, *see* Niobate				
Orthophosphate, *see* Phosphate				
Orthophosphite, *see* Phosphite				
Orthosilicate, *see* Silicate				
Orthovanadate, *see* Vanadate				
Oxide				
aluminum				319
αaluminum			140a	—
antimony				328, 329
cadmium				354
calcium				318
cerium (IV)				359
chromium (III)				336
cobalt				341, 342
copper(I)				344
copper (II)				345
dysprosium				361
erbium				363
germanium	143			325
hafnium				355
holmium				362
iodine				331
indium	141			320
iron (III) (hematite)				339
iron (magnetite)				340
lead				327
lithium				316
magnesium				317
manganese (II)				337
manganese (IV)				338
mercury (II)				358
molybdenum (IV)				351
molybdenum (VI)				352
nickel				343

Compound	Vols. 2–3 Spectrum Number			Vol. 4 Spectrum Number
	IR	R	R in water solution	
niobium (IV)				348
niobium (V)				349, 350
samarium				360
silicon				323
silicon (cristobalite)	142			322
silicon (vycor)				324
silver				353
tantalum	149	149		356
tellurium	144	144		330
thallium (III)				321
thorium				365
tin (II)				326
titanium (anatase)	145	145		332
titanium (rutile)	146	146		332
tungsten				357
uranium (IV)				366
uranium (orthorhombic)	150			367
uranium (hexagonal)				368
vanadium	147			333
vanadium	148			334
vanadium				335
ytterbium				364
yttrium				347
zinc				346
per-				
strontium		151		369
zinc		152		370
Oxychloride, *see* Chloride				
Oxyhydride, *see* Hydroxide				
Oxyiodide, *see* Iodide				
Palladate				
hexachloro-, potassium				775
tetrachloro-				
ammonium				772
potassium				774
sodium				773
Paramolybdate, *see* Molybdate				
Pentachloroferrate, *see* Ferrate				
Pentafluoroferrate, *see* Ferrate				
Pentafluorouranate, *see* Uranate				
Pentafluorozirconate, *see* Zirconate				
Pentanitrocerate (III), *see* Cerate (III)				
Perborate, *see* Borate				
Perchloroate, *see* Chlorate				
Periodate, *see* Iodate				
Permanganate, *see* Manganate				
Peroxide, *see* Oxide				

Compound	Vols. 2–3 Spectrum Number			Vol. 4 Spectrum Number
	IR	R	R in water solution	
Peroxydisulfate, *see* Sulfate				
Phosphate				
hexafluoro-				
ammonium		288		681
cesium				684
potassium	289	289	(289)	682
potassium and KHF$_2$				683
meta-				
aluminum				221
barium				220
beryllium	92	92		216
calcium	93	93		218
lead	95	95		222
magnesium				217
potassium	91	91		215
sodium	90	90		214
strontium	94	94		219
zinc				223
ortho-				
aluminum				240
ammonium (monobasic)				224
ammonium (dibasic)				226
ammonium cobalt	116			256
ammonium magnesium		115		254
ammonium manganese				255
ammonium sodium (dibasic)	98	98		—
antimony				243
barium (dibasic)		99		231
barium		103		—
bismuth	107	107		244
boron	104			239
cadmium		114		252
calcium				238
calcium (dibasic)				228
calcium nickel				258
chromium	108			245
cobalt				232
copper (II)	112			249
iron (II)	109			246
iron (III)	110			247
lead	106	106		242b
lead (monobasic)		97		—
lead (apatite)				242c
lead copper (I)	118			259
lithium		100		234
lithium (di)sodium	117			257
magnesium	102			236

Compound	Vols. 2–3 Spectrum Number			Vol. 4 Spectrum Number
	IR	R	R in water solution	
magnesium (basic)				237
mercury				253
nickel	111			248
potassium (monobasic)			(96a)	225
potassium (dibasic)				227
silver				251
sodium		101		235
sodium (monobasic)		96		—
strontium (dibasic, α-form)				229
strontium (dibasic, β-form)				230
tin (II)	105			241
zinc		113		250
pyro-				
aluminum				269
barium				271
barium (α-form)		121		270
calcium (β-form)				264
calcium (δ-form)	120			—
calcium (γ-form)				268
cobalt				274
copper (II)				276
diamyl ammonium		125		—
lead				273
magnesium				263
nickel				275
potassium				262
potassium sodium				267
sodium	119			260, 261
sodium (dibasic)				233
strontium (α-form)				265
strontium (β-form)				266
tin		123		272
Pyrophosphate				
titanium		122		—
zinc				277
tripoly-				
potassium				279
sodium	124	124		278
Phosphide				
antimony				200
bismuth				201
zinc				203
Phosphite				
hypo-				
ammonium				204
barium	88	88		—
calcium	86	86		208

Compound	Vols. 2–3 Spectrum Number			Vol. 4 Spectrum Number
	IR	R	R in water solution	
iron (II)				204
iron (III)				210
lithium				205
magnesium	87	87		—
manganese				209
potassium		85	(85)	207
sodium	84	84		206
ortho-				
barium				212
sodium				211
Phosphoric acid, meta-				
Phosphorodifluorodate, *see* Flurodate				
Phosphorofluoroidate, *see* Fluoride				
Phosphoromolybdate, *see* Molybdate				
Phosphorothioate, *see* Thioate				
Phosphorotungstate, *see* Tungstate				
Platinate				
cyani, barium				43
tetrachloro-				
ammonium				777
potassium				778
Pyroarsenate, *see* Arsenate				
Pyrophosphate, *see* Phosphate				
Pyrosulfate, *see* Sulfate				
Pyrosulfite, *see* Sulfite				
Pyrovanadate, *see* Vanadate				
Selenate				
ammonium		224		522
calcium	225	225		256
copper (II)		226		529
iron (II)				527
magnesium				525
nickel				528
potassium				524
potassium alumino	229	229		—
silver	228	228		520
sodium			(224a)	523
zinc		227		—
Selenide				
chormium				510
gallium				506
lead	220			508
molybdenum				514
niobium				513
tantalum				515
tin (II)	219			507
titanium	221			509

Compound	Vols. 2–3 Spectrum Number			Vol. 4 Spectrum Number
	IR	R	R in water solution	
tungsten				516
zinc				511
zirconium				512
Selenite				
barium				519
copper	223			521
potassium				518
sodium		(221a)		517
zinc	222			520
Silica gel				94
Silicate				
barium zirconium	47			97
cobalt	46			—
lithium				87
lithium zirconium				98
magnesium aluminum				99
magnesium calcium aluminum				96
hexafluoro-				
ammonium		284		653
barium				659
calcium				658
cobalt				661
copper (II)				663
lithium				654
magnesium				657
manganese				660
nickel				662
potassium				656
sodium		285		655
zinc				664
meta-, lithium				88
ortho-				
cobalt				91
copper (II)				92
lead				90
magnesium				89
zinc				93
Silicide				
boron	41			78
calcium	40			77
magnesium	39			76
manganese	44			81
molybdenum	45			84, 85
niobium				83
titanium	42			79
tungsten				86
vanadium	43			80

Compound	Vols. 2–3 Spectrum Number			Vol. 4 Spectrum Number
	IR	R	R in water solution	
zirconium				82
Stannate (IV)				
barium				125
bismuth				127
calcium				123
cerium				129
iron				128
lead				126
magnesium				122
strontium				124
hexachloro-				
ammonium				765
cobalt				766
hexafluoro-				
calcium				677
cobalt				678
copper (II)				680
lithium				673
magnesium				676
nickel				679
potassium				675
sodium				674
hexahydroxo-				
cadmium	159			382
copper (II)	157			380
potassium				379
zinc	158			381
trichloro-				
ammonium				763
potassium				764
trifluoro-				
ammonium				668
iron (II)				671
potasisum				670
sodium				669
zinc				672
Subnitrate, *see* Nitrate				
Sulfamate, lead				412
Sulfate				
aluminum	193	193		444
aluminum sodium	214	214		—
ammonium				431
ammonium antimony				
trifluoride complex				482
ammonium cadmium	216	216		—
ammonium chromium				483
ammonium cobalt	212			488

Compound	Vols. 2–3 Spectrum Number			Vol. 4 Spectrum Number
	IR	R	R in water solution	
ammonium copper (II)				489
ammonium hydrogen				414
ammonium imidodi-				413
ammonium iron (II)				485
ammonium iron (III)	210	210		486, 487
ammonium manganese	209	209		484
ammonium sodium				481
antimony		196		449
barium	192	192		443
beryllium		188		437
bismuth	197	197		450
cadmium	205	205		463
calcium	190	190		440, 441
cerium (III)	206	206		467
cerium (IV)	207	207		468
cesium				436
cesium aluminum	215	215		499
cobalt (II)	200	200		455
copper (II)	201			457
copper tetraamine				480
dysprosium				474
erbium				476
europium				472
gadolinium				473
gallium				445
holmium				475
indium				446
iron (II)				453
iron (III)		199		454
lead		195		448
lithium	185	185		432
magnesium	189	189		438, 439
manganese (II)				452
mercury (I)				464
mercury (II)				465
neodymium				470
nickel				456
potassium	187	187		434
potassium aluminum				492
potassium cadmium				497
potassium chromium		213		493
potassium copper (II)				496
potassium hydrogen		170		416
potassium iron (III)				494
potassium magnesium				491
potassium nickel				495
praseodymium				469

Compound	Vols. 2–3 Spectrum Number			Vol. 4 Spectrum Number
	IR	R	R in water solution	
rubidium				435
rubidium aluminum				498
rubidium hydrogen				417
samarium				471
silver	204	204		462
sodium	186	186		433
sodium hydrogen		169	(169)	415
sodium iron (III)				490
strontium	191	191		442
thallium	194	194		447
thorium	208	208		478
uranium				479
vanadium	198	198		451
ytterbium				477
ytterium				460
zinc	202	202		458, 459
zirconium	203	203		461
imidodisulfate				
ammonium	168	168		413
peroxydi-				
ammonium				501
potassium	217	217		503
sodium	216a	216a	(216a)	502
pyro-, silver				430
thio-				
barium	172	172	(172)	420
lead	173			421
magnesium				419
potassium	171	171	(171)	418
Sulfide				
antimony	161			397
arsenic (IV)	160			395
arsenic (V)				396
bismuth	162			398
cadmium				407
copper				402
mercury (II)				411
molybdenum				405
nickel	165			401
niobium				404
silver	166			406
tantalum (II)				408
tantalum (IV)	167			409
tellurium	163			399
titanium	164			400
tungsten				410
zinc				403

Compound	Vols. 2–3 Spectrum Number			Vol. 4 Spectrum Number
	IR	R	R in water solution	
Sulfite				
ammonium		176		—
barium	181	181		427
lead				428
magnesium	179	179		425
potassium		178		—
sodium		177		424
strontium		180		426
pyro-				
potassium		175		423
sodium	174	174		422
Sulfonate				
fluoro-				
ammonium				514
potassium				505
Sulfur				411a
Tantalate				
heptafluoro-, potassium	300	300		713
hexafluoro-, potassium				712
Telluric acid				539
Telluride				
bismuth				532
chromium				535
molybdenum				537
tin (II)	230			531
titanium				533
tungsten				538
vanadium				534
zinc	230a			536
Tetraborate, *see* Borate				
Tetrachlorcuprate, *see* Cuprate				
Tetrachloropalladate, *see* Palladate				
Tetrachloroplatinate, *see* Platinate				
Tetrafluoroaluminate, *see* Aluminate				
Tetrafluoroantimonate, *see* Antimonate				
Tetrafluoroborate, *see* Borate				
Tetraflorozirconate, *see* Zirconate				
Tetrahydroaluminate, *see* Aluminate				
Tetraiodomercurate, *see* Mercurate				
Tetrathiotungtate, *see* Tungstate				
Thiocarbonate, *see* Carbonate				
Thiocyanate, *see* Cyanate				
Thioate				
phosphoro-, sodium				280
Thionate				
di, potassium				429
di, sodium		182		—

Compound	Vols. 2–3 Spectrum Number			Vol. 4 Spectrum Number
	IR	R	R in water solution	
Thiosulfate, *see* Sulfate				
Titanate(IV)				
barium	49	49		103
bismuth				105
calcium				101
cerium				110
cobalt				106
copper				108
europium				111
lead				104
lithium				100
nickel				107
strontium		48		102
zinc				109
hexafluoro-				
ammonium				689
barium				694
calcium				693
lithium				690
nickel				695
potassium	291			692
sodium				691
Trichlorostannate, *see* Stannate				
Trifluorostannate, *see* Stannate				
Tripolyphosphate, *see* Phosphate				
Tungstate				
aluminum				585
ammonium				592
ammonium	255	255		—
barium				584
cadmium				590
calcium	257	257		582
chromium				593
copper				586
lithium				578
magnesium				581
potassium				580
potassium	262			—
para, sodium	261	261,261		591
silver	260			589
sodium	256	256	(256)	579
strontium				583
zinc	258	258		587
zirconium	259			588
boro-, barium	263	263		595
pospho-				
ammonium				596

Compound	Vols. 2–3 Spectrum Number			Vol. 4 Spectrum Number
	IR	R	R in water solution	
sodium				597
tetrathio-, ammonium				594
Uranate				
ammonium				882
lead calcium				884
sodium				883
pentafluoro-, sodium	301			714
Vanadate				
meta-, ammonium	137	137		305
sodium	138			—
ortho-				
calcium				308
calcium copper (II) hydroxy-				313
calcium nickel hydroxy-				312
iron (III)				310
lead				309
magnesium		140		—
silver				311
sodium				307
pyro-, sodium			(138a)	306
lead	139			—
zincate tetra fluoro, potassium	292			703
Zirconate				
barium				116
bismuth				118
cadmium				120
calcium				114
cerium	53	53		121
lead	51	51		117
lithium				112
magnesium				113
strontium	50			115
zinc	52			119
heptafluoro-, potassium				710
hexafluoro-				
ammonium	295	295		706
indium	298			709
potassium				708
sodium	296	296		707
oxo-, potassium		303		—
pentafluoro-				
potassium	294			705
sodium	293	293		704

B. ALPHABETICAL INDEX OF SPECTRA FOR NONIONIC COMPOUNDS AND HYDRAZINE SALTS

	Spectrum Number		Vol. 4
Compound	IR	R	
ammonia	381		
ammonia-d$_3$	382		
arsine	408		
boron tribromide	368		
boron trifluoride	367		
bromine		430	
carbon dioxide	379		
carbon disulfide	372	372	
carbon monoxide	369		
carbonyl chloride	380		
carbonyl fluoride	370		
carbonyl sulfide	371		
chlorine		429	
chloromethylphosphonic dichloride	407		
cyanamide	377		
cyanogen	374		
cyanogen bromide	376		
cyanogen chloride	375		
cyanoquanidine	378		
decafluorodisulfide	426		
decaborane	364		
decaborane monohydrazine	366		
decafluorodisulfide	426		
deuterium bromide	435		
deuterium chloride	433		
deuterium oxide	411		
diborane	363		
hydrazine diborane	365		
hydrazine dichloride	389		
hydrazine tetrafluoride	387		
hydrogen bromide	434		
hydrogen chloride	432		
hydrogen cyanide	373		
hydrogen sulfide	413		
iodine		431	
methyl chlorosulfinate	427	427	
nitric oxide	384		
nitrogen trichloride	386		
nitrogen trifluoride	385		
nitrous oxide	383		
ozone	409		
phosphine	390		
phophoric acid	404		
phophorous tribromide	393		
phosphorus trichloride	392		
phosphorus trifluoride	391		
phosphorus triiodide	394		

Compound	Spectrum Number		Vol. 4
	IR	R	
phosphorus oxybromide	397		
phosphorus oxychloride	396		
phosphorus oxyfluoride	395		
phosphorus pentabromide	402		
phosphorus pentachloride	401		
phosphorus pentasulfide	406		
phosphorus thioamide	405		
potassium methylsulfate	428	428	
pyrosulfuryl chloride	421		
sulfonamide	423		
β-sulfonyl chloride	422		
sulfur	412	412	811a
sulfur dichloride	416		
sulfur dioxide	414		
sulfur hexafluoride	425		
sulfuric acid	424		
sulfur monobromide	417		
sulfur monochloride	415		
sulfuryl chloride	420		
thionyl chloride	419	419	
thionyl fluoride	418		
thiophosphoryl bromide	400		
thiophosphoryl chloride	399		
thiophosphoryl dichloride fluoride	398	398	
trichlorophosphazosulfuryl chloride	403		
water	410		
water-D_2O	411		

C. ALPHABETICAL INDEX OF MISCELLANEOUS MINERALS

Spectrum Number	Compound	Formula	Spectra IR	R	Vol. 4 No. IR
436	Aragonite	$CaCO_3$	X	X	
437	Asbestos (amphibole)		X		
438	Asbestos (crocidolite)		X		
439	Asbestos (serpentine chrysolite)		X		
440	Gearksutite	$CaAlF(OH)$	X	X	
441	Hydromagnesite	$3Mg \cdot Mg(OH)_2 \cdot 3H_2O$	X	X	
442	Itacolumite		X	X	
443	Kalinite	$AlK(SO_4)_2 \cdot 12H_2O$	X	X	
444	Meyerhofferite	$2Ca \cdot 3B_2O_3 \cdot 7H_2O$	X	X	
445	Quartz	SiO_2	X	X	892
446	Realgar	AsS	X	X	
447	Stilbite	$H_4(Na_2,Ca)Al_2Si_6O_{18} \cdot 4H_2O$	X	X	
448	Thaumasite	$CaSiO_3 \cdot CaCO_3 \cdot CaSO_4 \cdot 15H_2OX$	X	X	
449	Vermiculite		X		

D. ALPHABETICAL INDEX OF SPECTRA FOR ORGANIC SALTS

Compound	OS Spectrum Number R	IR
Acetate		
barium	11	11
calcium	9	9
guanidinium	13	13
sodium	8	8
strontium	10	10
thallous	12	12
2-cyano-, sodium	20	20
Benzenesulfonate		
sodium	68	68
4-hydroxy-, sodium	69	69
Benzoate		
bismuth	52	52
calcium	51	51
lead	55	55
lithium	49	49
manganese	53	53
zinc	54	54
Butyrate		
calcium	15	15
sodium	14	14
Cinnamate		
sodium	57	57
Citrate		
bismuth	41	41
bismuth ammonium	43	43
calcium	40	40
lithium	38	38
manganese	42	42
manganese sodium	44	44
sodium	39	39
Cyanoacetate		
sodium	20	20
Ethylenediaminetetraacetic acid		
stannous	45	45
Formate		
barium	5	5
calcium	3	3
lead	6	6
magnesium	2	2
sodium	1	1
strontium	4	4
zinc	7	7
Hippurate		
calcium	48	48
lithium	46	46
sodium	47	47

E. ALPHABETICAL INDEX FOR INFRARED WINDOWS

W

Window Material	IR Spectrum Number
barium fluoride	11
cadium teluride	10
calcium fluoride	13
lithium fluoride	5
potassium bromide	7
potassium chloride	14
pyrex	1
quartz	2
silicon carbide	4
silicon wafer	3
silver chloride	8
sodium chloride	6, 12
zinc selinide	9

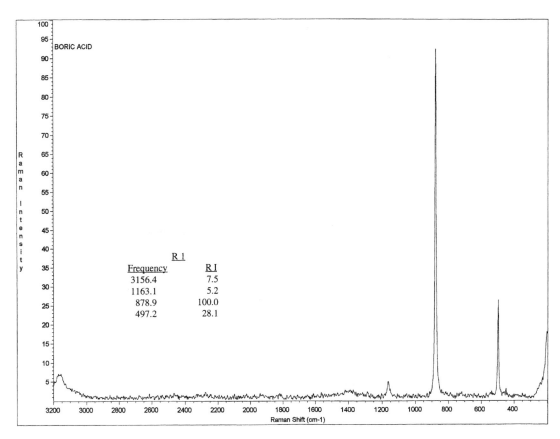

1   Boric acid H_3BO_3

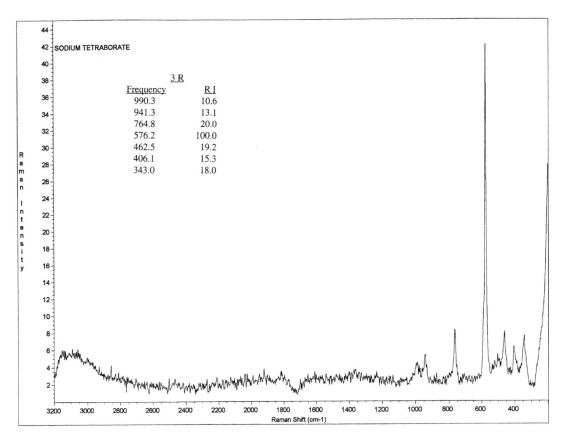

3   Sodium tetraborate $Na_2B_4O_7 \cdot 5H_2O$

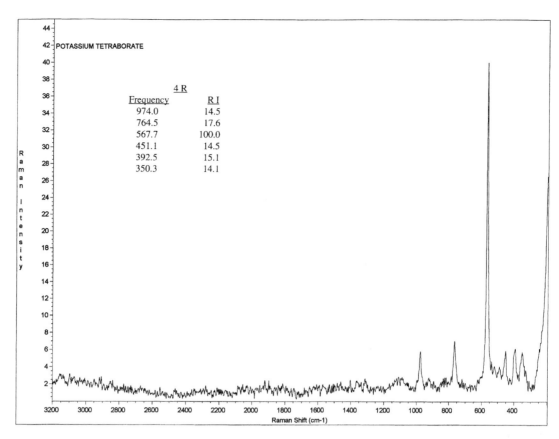

4   Potassium tetraborate $K_2B_4O_7 \cdot 8H_2O$

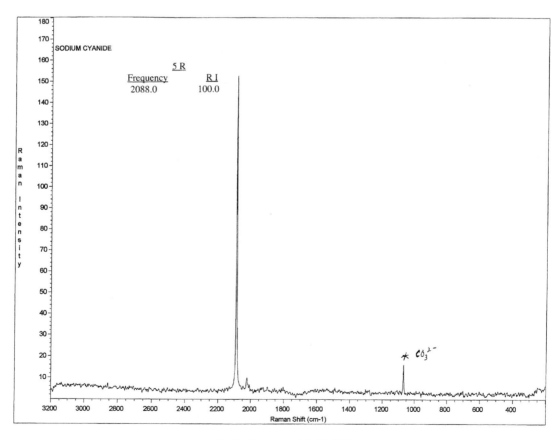

5   Sodium cyanide NaCN

45

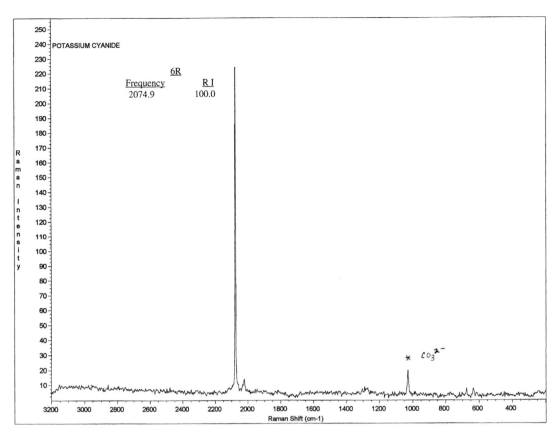

6   Potassium cyanide KCN

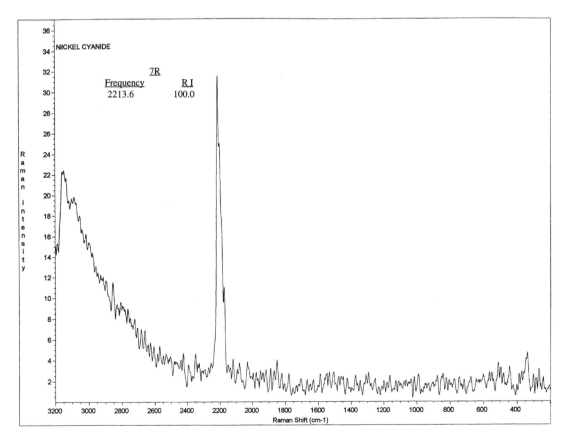

7   Nickel (II) cyanide Ni(CN)$_2$·4H$_2$O

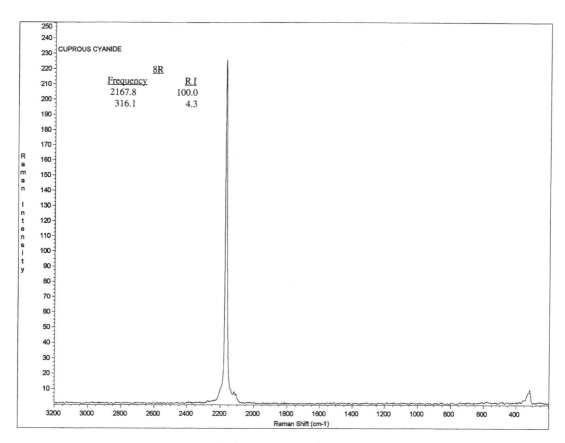

8   Copper (I) cyanide CuCN

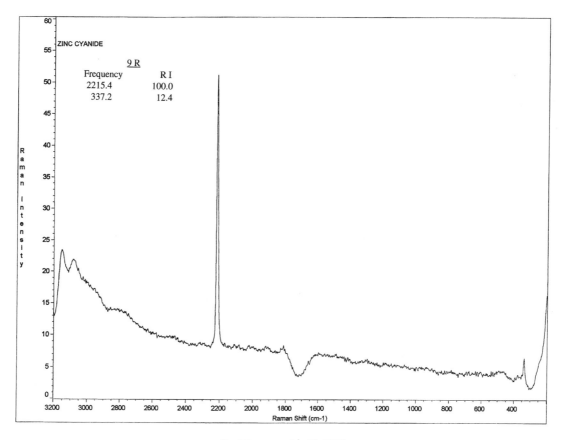

9   Zinc cyanide Zn(CN)$_2$

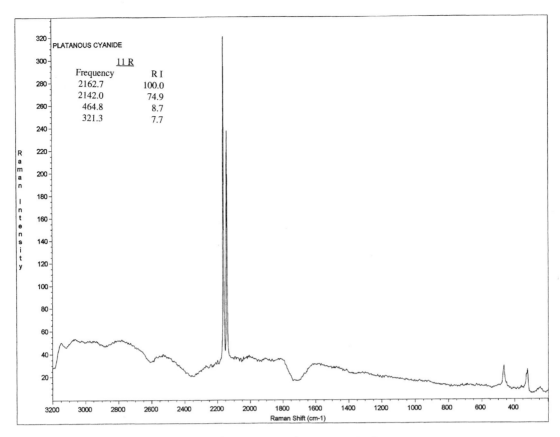

11   Platinous cyanide Pt(CN)$_2 \cdot$xH$_2$O

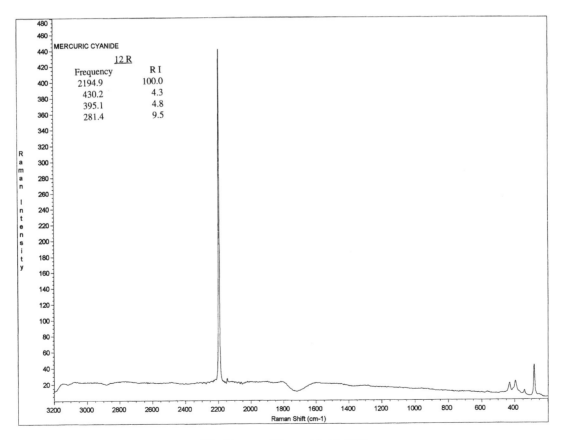

12   Mercury (II) cyanide Hg(CN)$_2$

48

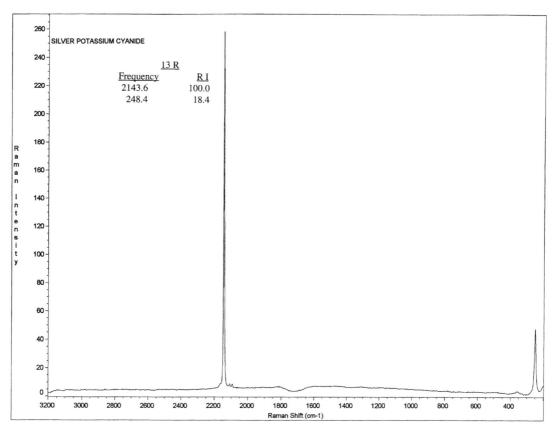

13   Potassium cyanoargenate KAg(CN)$_2$

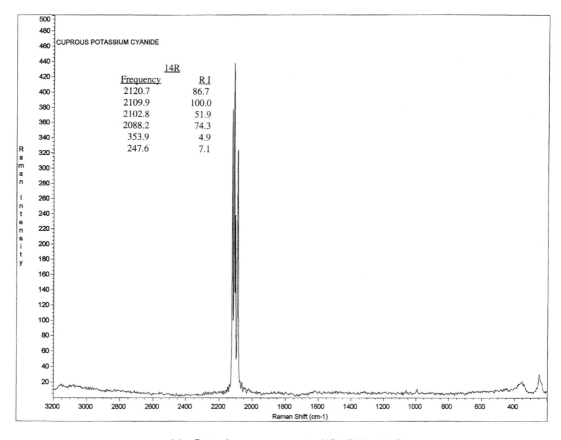

14   Potassium cyanocuprate KCu(CN)$_2 \cdot$xH$_2$O

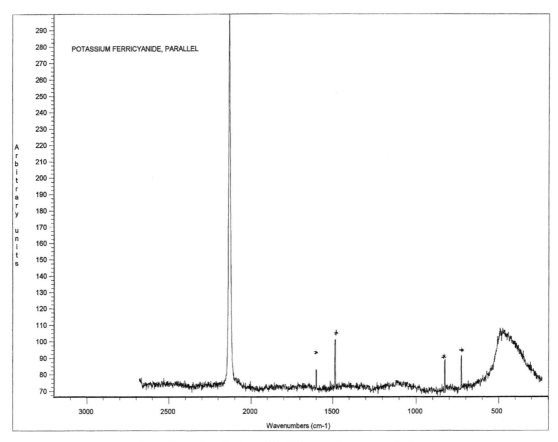

14a   Potassium ferricyanide $K_3Fe(CN)_6$ in water solution

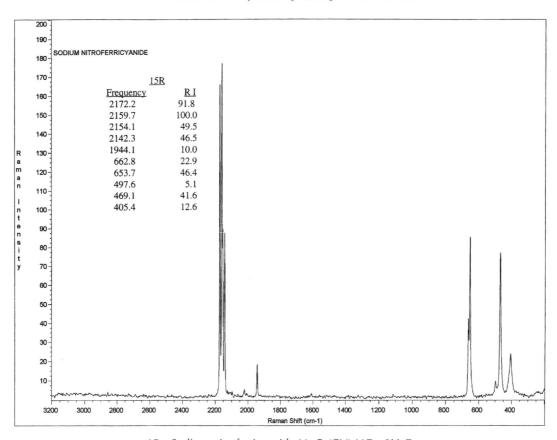

15R	
Frequency	R I
2172.2	91.8
2159.7	100.0
2154.1	49.5
2142.3	46.5
1944.1	10.0
662.8	22.9
653.7	46.4
497.6	5.1
469.1	41.6
405.4	12.6

15   Sodium nitroferricyanide $Na_2Fe(CN)_5NO_2 \cdot 2H_2O$

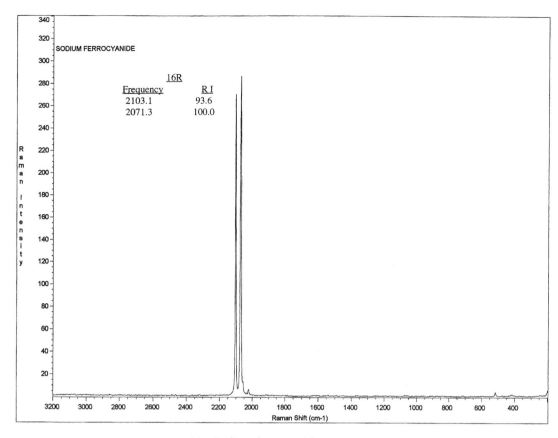

16   Sodium ferrocyanide Na$_4$Fe(CN)$_6$

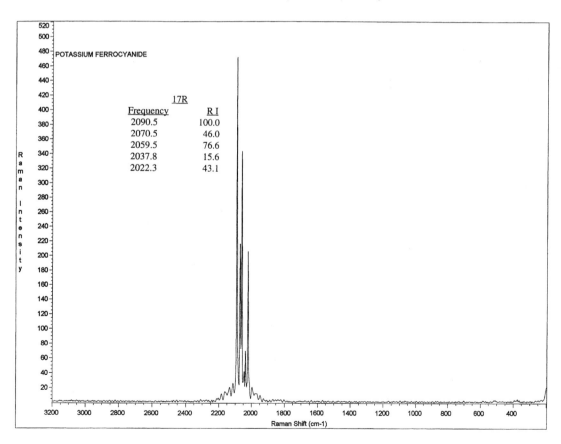

17   Potassium ferrocyanide K$_4$Fe(CN)$_6$

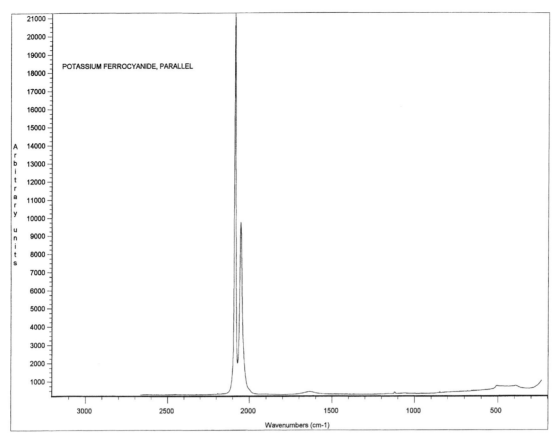

17a   Potassium ferrocyanide $K_4Fe(CN)_6$ in water solution

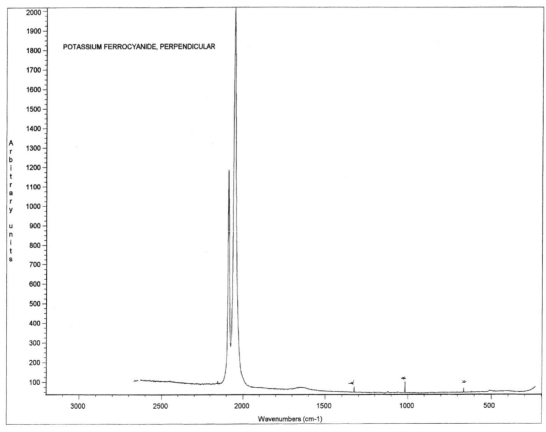

17b   Potassium ferrocyanide $K_4Fe(CN)_6$ in water solution

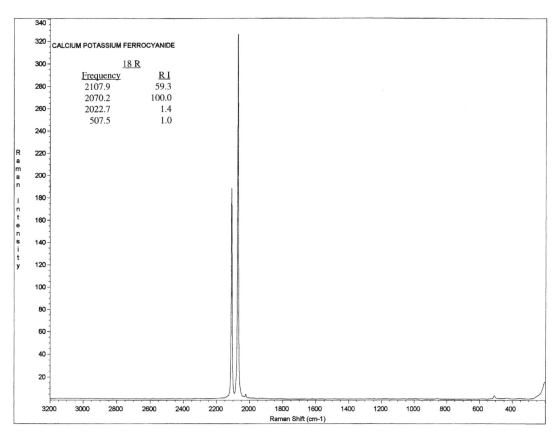

18   Potassium calcium ferrocyanide K$_2$CaFe(CN)$_6$

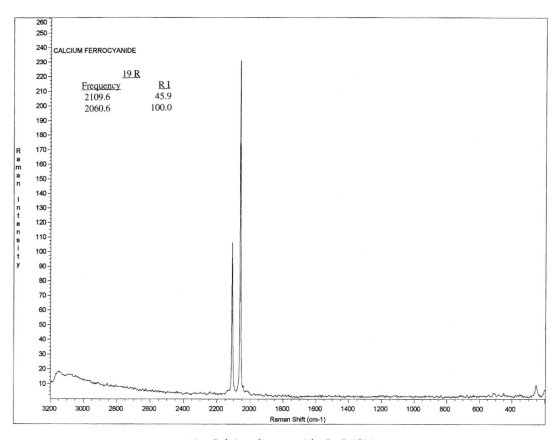

19   Calcium ferrocyanide Ca$_2$Fe(CN)$_6$

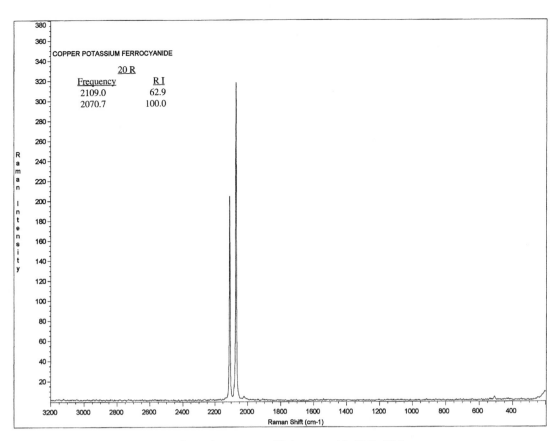

20   Potassium copper (II) ferrocyanide $K_2Cu(CN)_6$

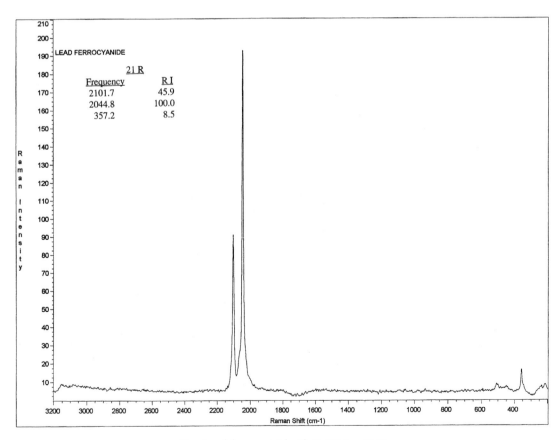

21   Lead ferrocyanide $Pb_2Fe(CN)_6 \cdot xH_2O$

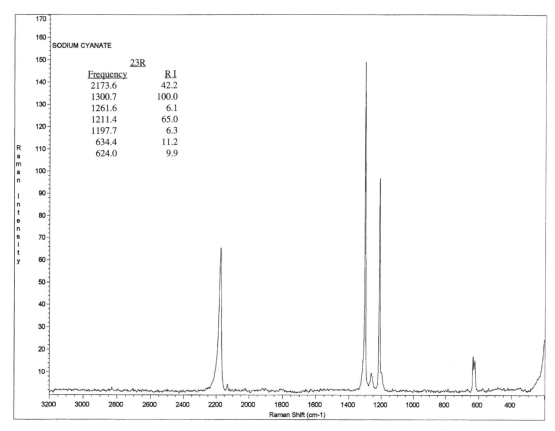

SODIUM CYANATE

23R	
Frequency	R I
2173.6	42.2
1300.7	100.0
1261.6	6.1
1211.4	65.0
1197.7	6.3
634.4	11.2
624.0	9.9

23   Sodium cyanate NaOCN

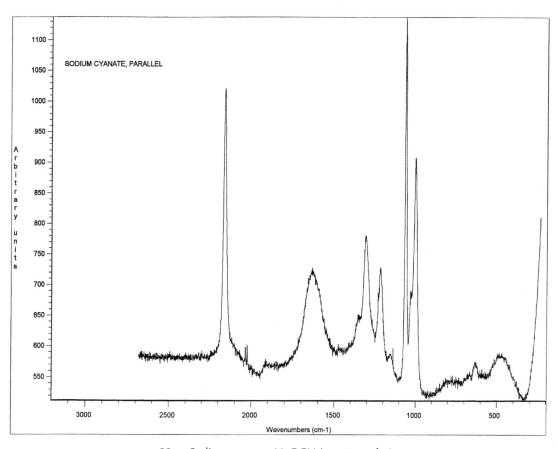

SODIUM CYANATE, PARALLEL

23a   Sodium cyanate NaOCN in water solution

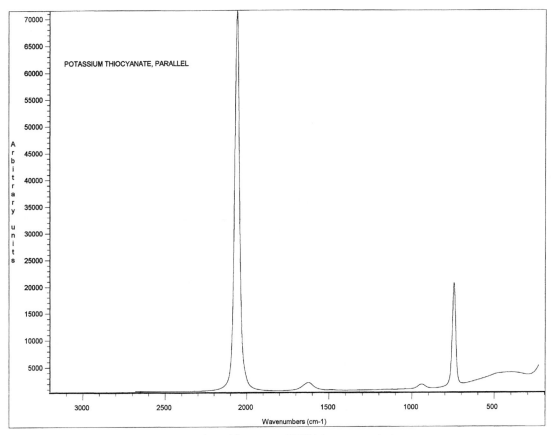

24   Potassium thiocyanate KSCN in water solution

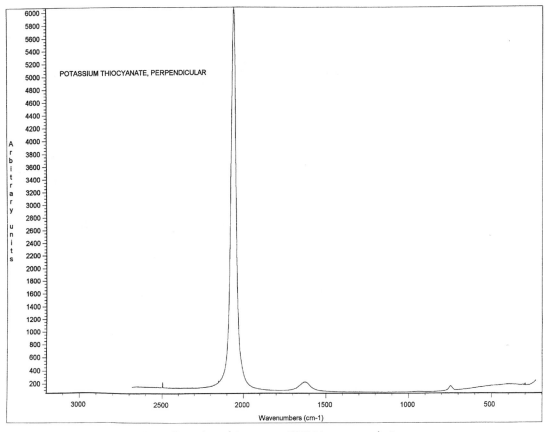

24a   Potassium thiocyanate KSCN in water solution

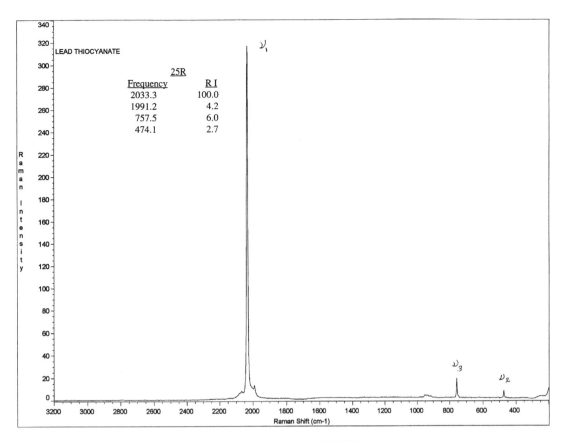

25   Lead thiocyanate Pb(SCN)$_2$

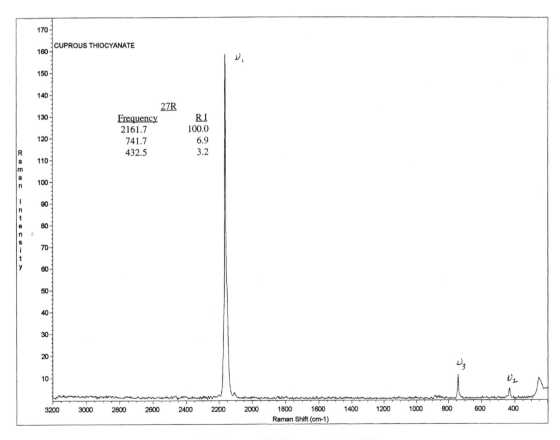

27   Copper (I) thiocyanate CuSCN

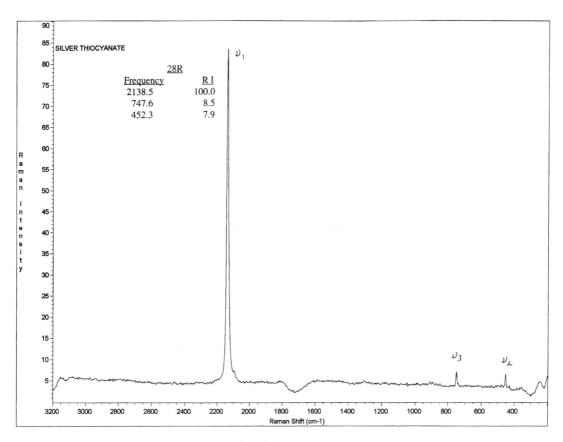

28 Silver thiocyanate AgSCN

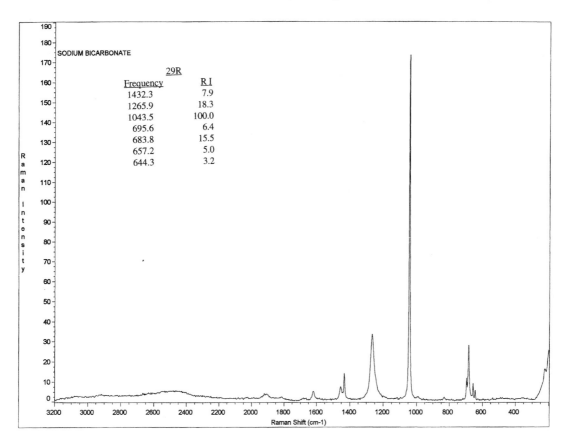

29 Sodium bicarbonate NaHCO$_3$

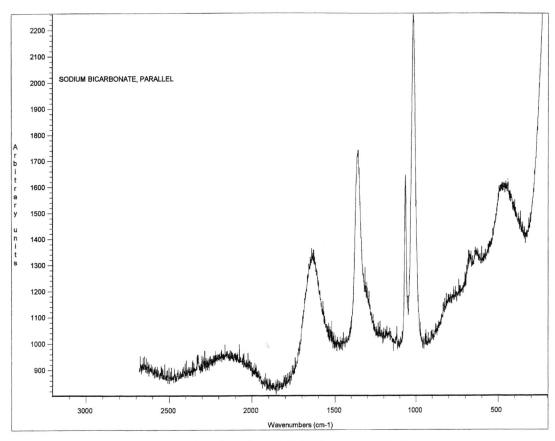

29a   Sodium bicarbonate NaHCO₃ in water solution

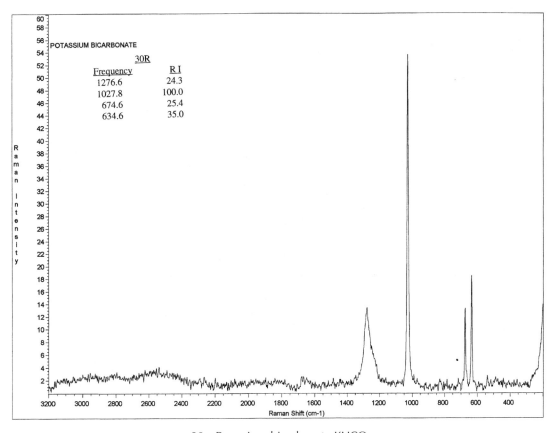

30   Potassium bicarbonate KHCO₃

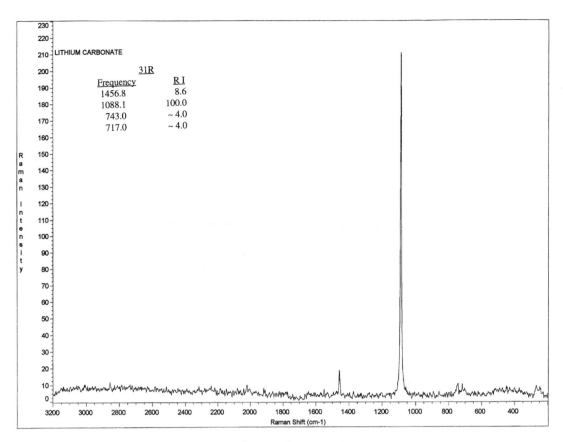

31 Lithium carbonate Li$_2$CO$_3$

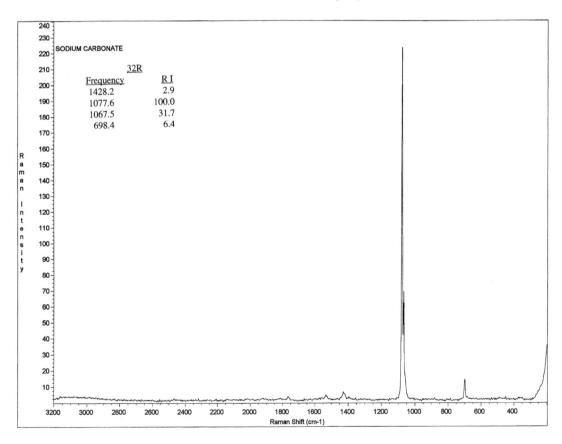

32 Sodium carbonate Na$_2$CO$_3$

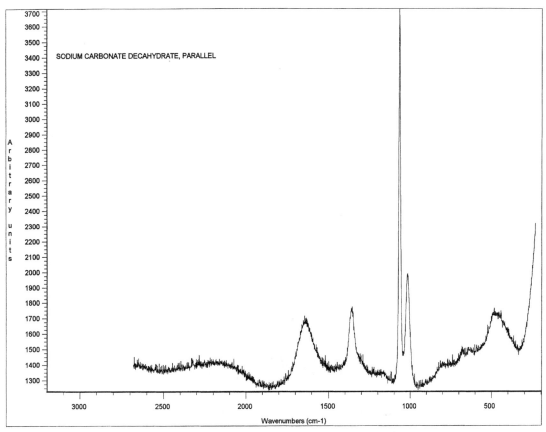

32   Sodium carbonate Na$_2$CO$_3$ in water solution

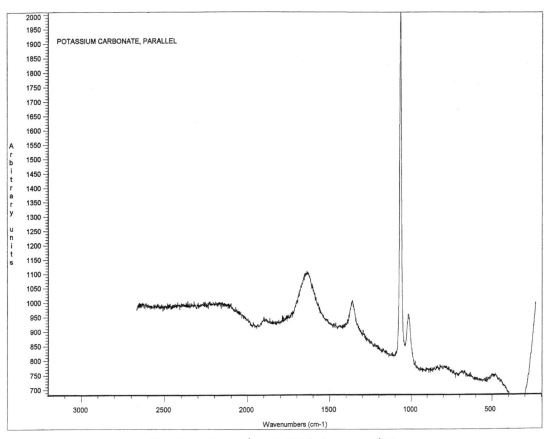

32a   Potassium carbonate K$_2$CO$_3$ in water solution

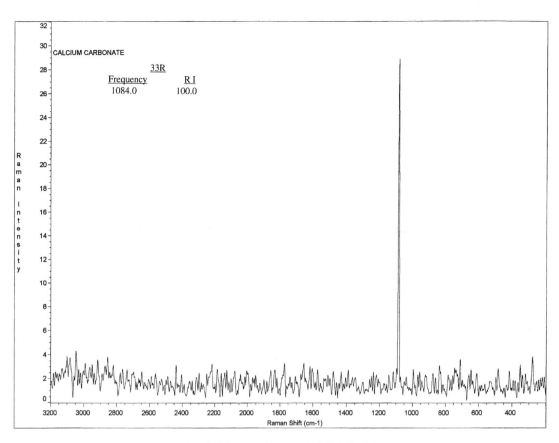

33   Calcium carbonate (calcite) CaCO$_3$

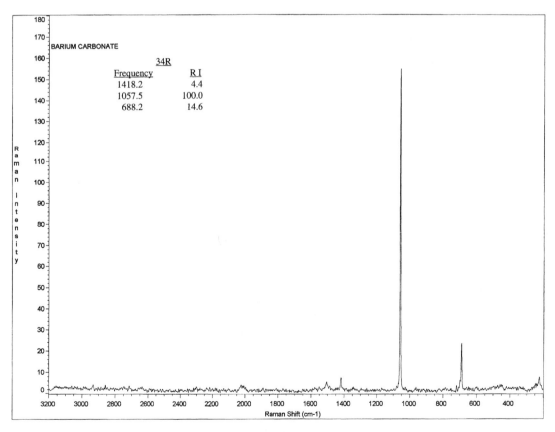

34   Barium carbonate BaCO$_3$

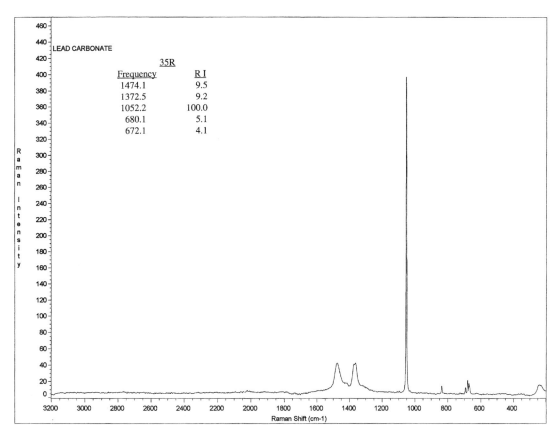

LEAD CARBONATE

35R

Frequency	R I
1474.1	9.5
1372.5	9.2
1052.2	100.0
680.1	5.1
672.1	4.1

35   Lead carbonate $PbCO_3$

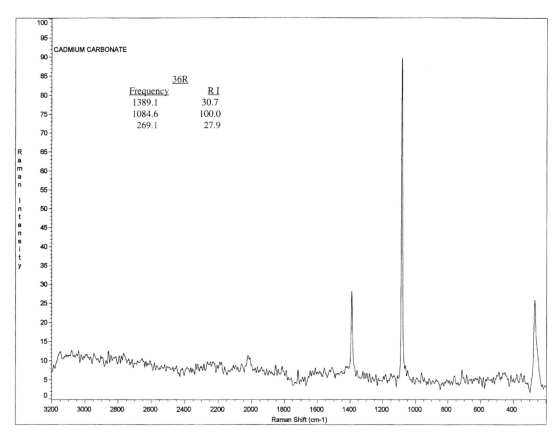

CADMIUM CARBONATE

36R

Frequency	R I
1389.1	30.7
1084.6	100.0
269.1	27.9

36   Cadmium carbonate $CdCO_3$

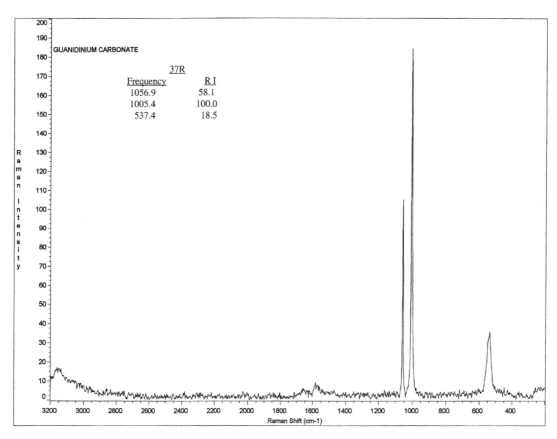

GUANIDINIUM CARBONATE

37R

Frequency	R I
1056.9	58.1
1005.4	100.0
537.4	18.5

37   Guanidinium carbonate [(H$_2$N$_2$)C$=$NH$_2$]$_2$CO$_3$

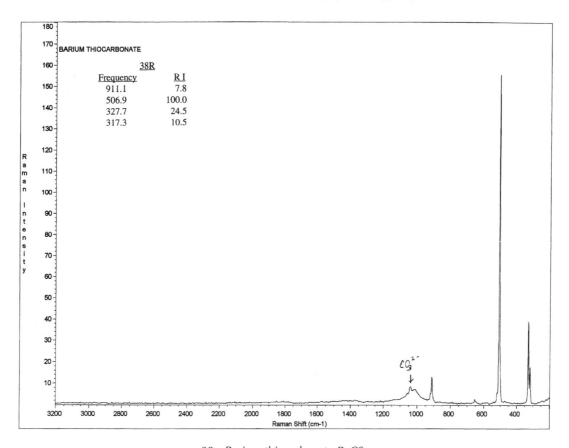

BARIUM THIOCARBONATE

38R

Frequency	R I
911.1	7.8
506.9	100.0
327.7	24.5
317.3	10.5

38   Barium thiocarbonate BaCS$_3$

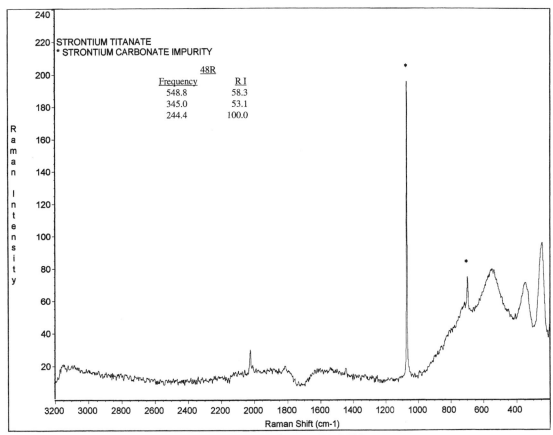

48   Strontium titanate (IV) SrTiO₃

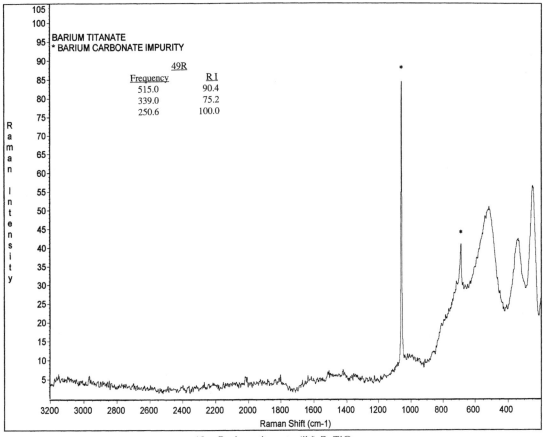

49   Barium titanate (IV) BaTiO₃

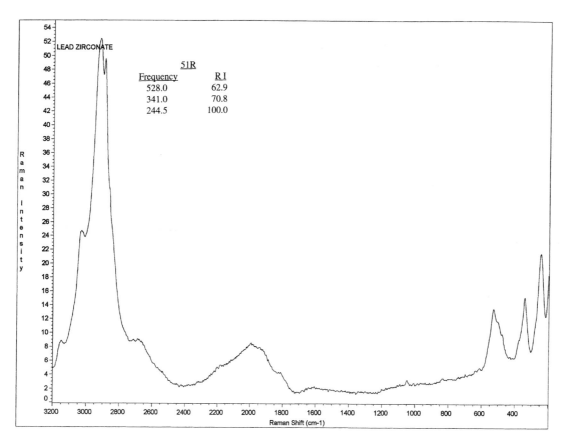

51   Lead zirconate (IV) PbZrO$_3$

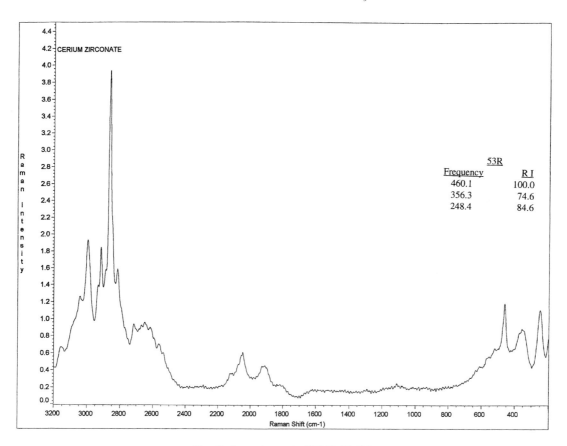

53   Cerium zirconate (IV) Ce(ZrO$_3$)$_2$

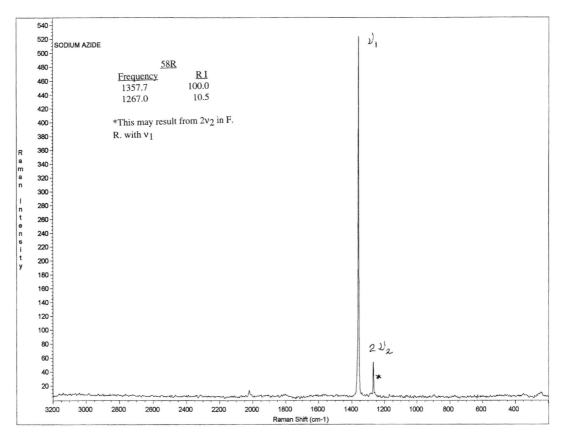

58   Sodium azide NaN₃

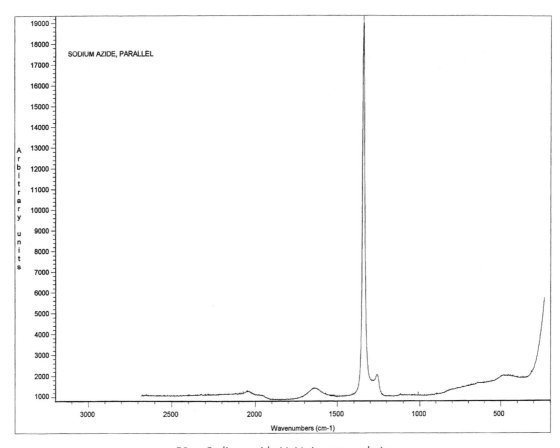

58a   Sodium azide NaN₃ in water solution

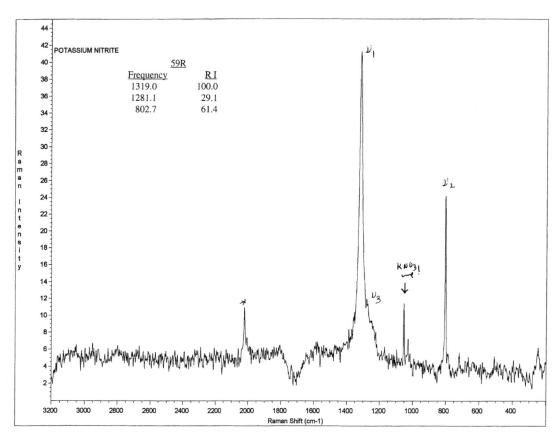

59 Potassium nitrite KNO$_2$·xH$_2$O

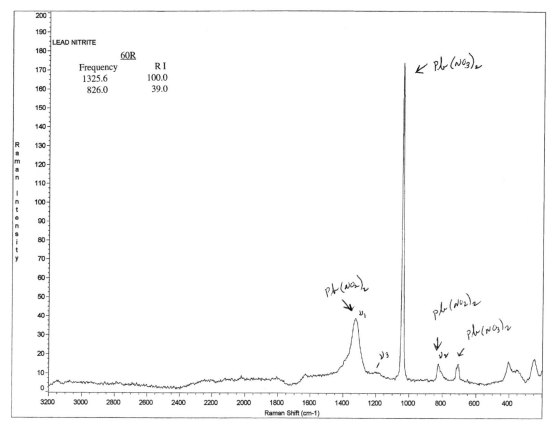

60 Lead nitrite Pb(NO$_2$)$_2$·xH$_2$O

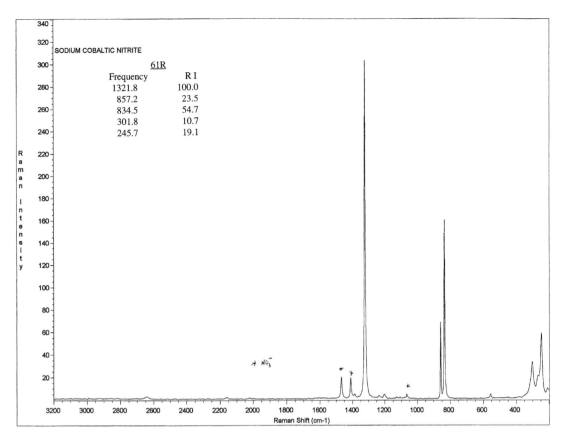

61   Sodium hexanitrocobaltate (III) Na$_3$Co(NO$_2$)$_6$

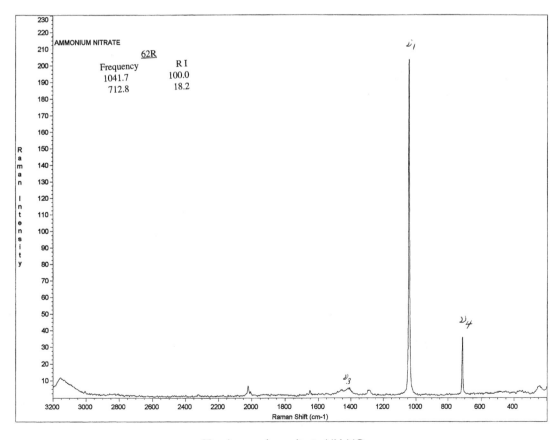

62   Ammonium nitrate NH$_4$NO$_3$

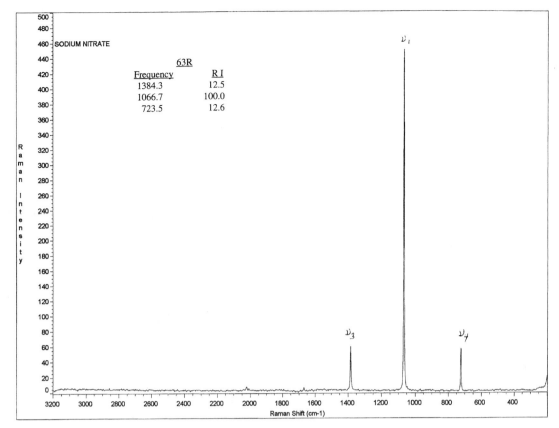

63   Sodium nitrate NaNO$_3$

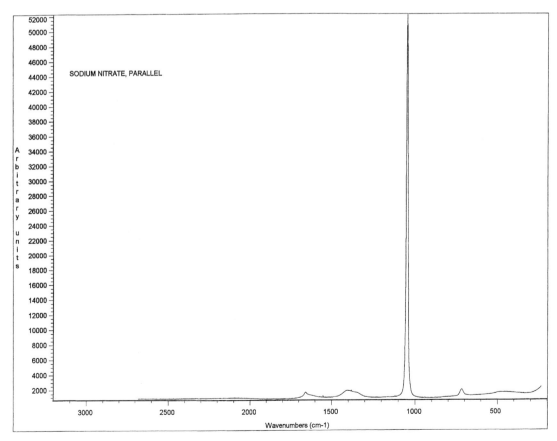

63a   Sodium nitrate NaNO$_3$ in water solution

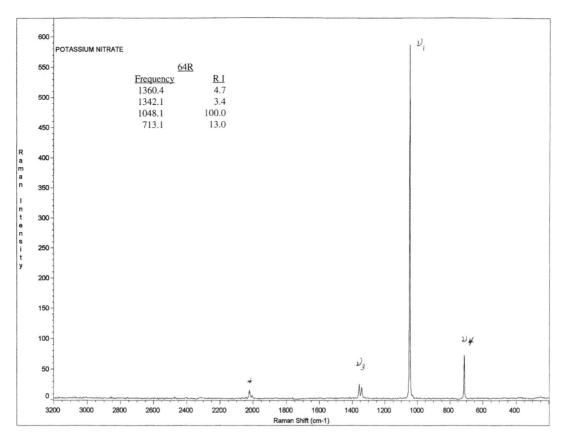

POTASSIUM NITRATE

64R

Frequency	R I
1360.4	4.7
1342.1	3.4
1048.1	100.0
713.1	13.0

64   Potassium nitrate KNO_3

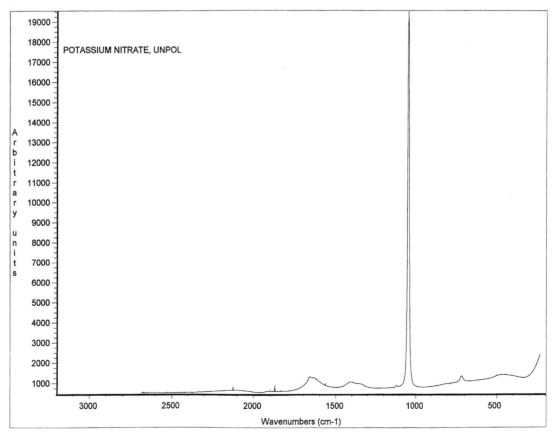

POTASSIUM NITRATE, UNPOL

64a   Potassium nitrate KNO_3 in water solution

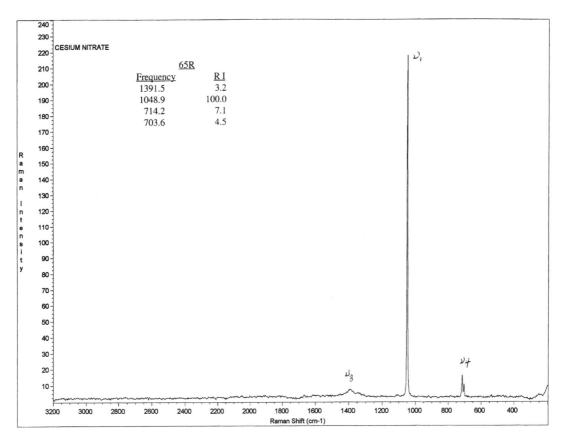

65   Cesium nitrate CsNO$_3$

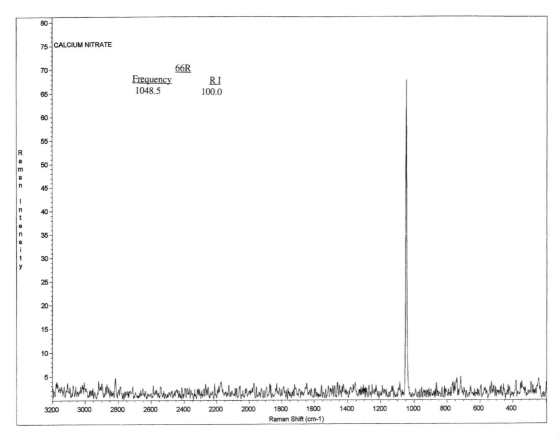

66   Calcium nitrate Ca(NO$_3$)$_2 \cdot$4H$_2$O

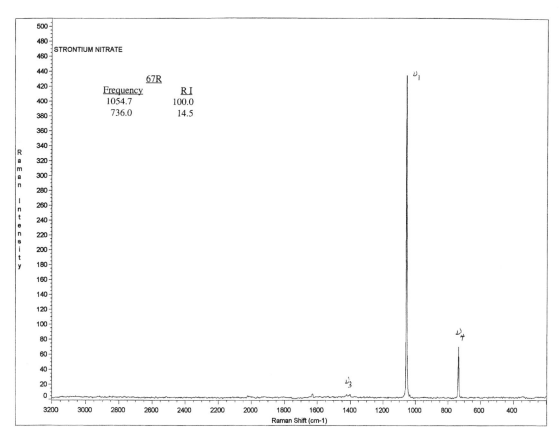

67   Strontium nitrate $Sr(NO_3)_2$

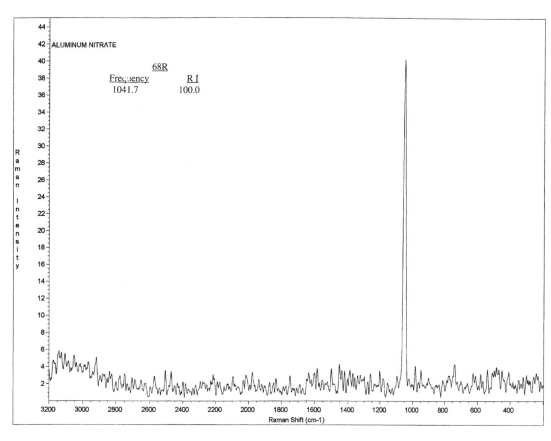

68   Aluminum nitrate $Al(NO_3)_3 \cdot 9H_2O$

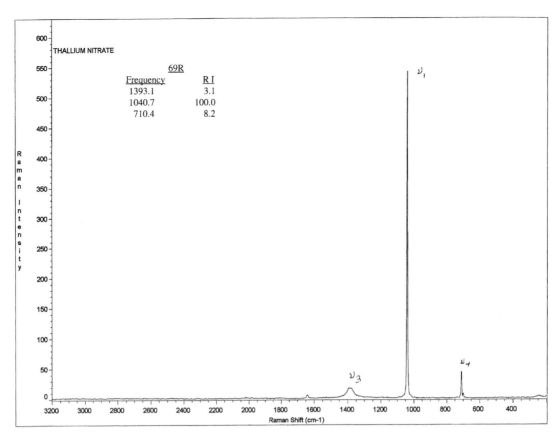

69   Thallium nitrate Tl(NO$_3$)$_3$

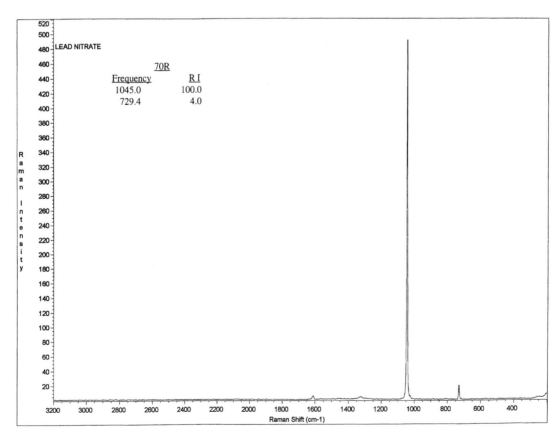

70   Lead nitrate Pb(NO$_3$)$_2$

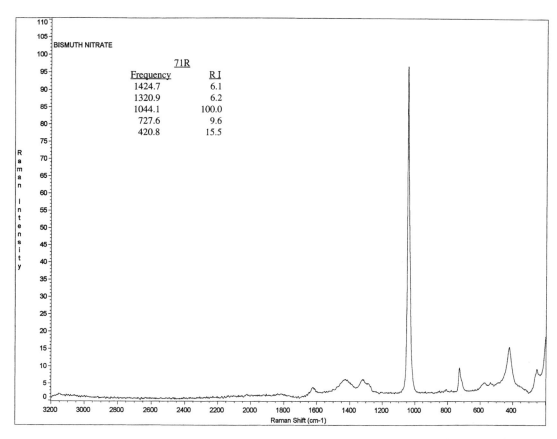

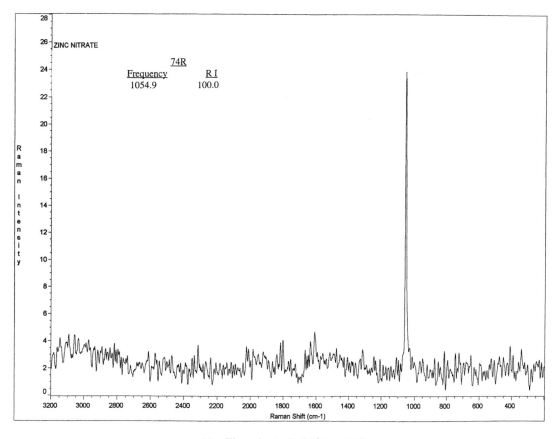

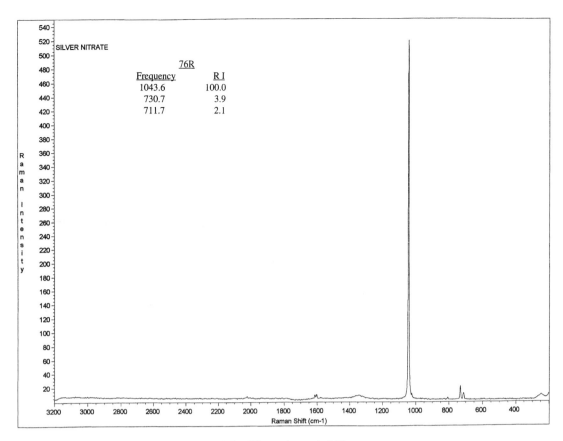

76 Silver nitrate AgNO$_3$

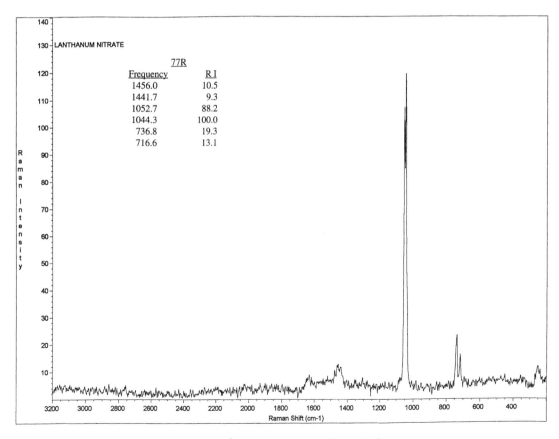

77 Lanthanum nitrate La(NO$_3$)$_3 \cdot$6H$_2$O

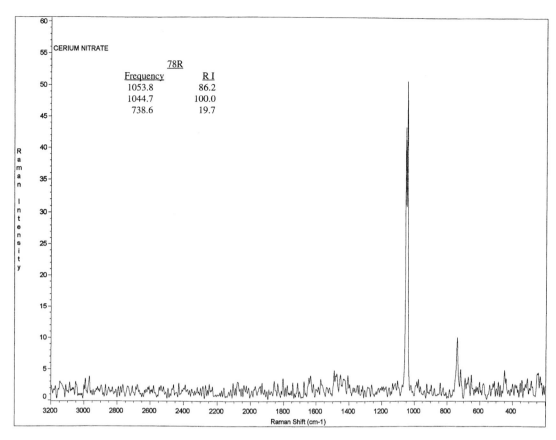

CERIUM NITRATE

78R

Frequency	R I
1053.8	86.2
1044.7	100.0
738.6	19.7

78 Cerium nitrate $Ce(NO_3)_3 \cdot 6H_2O$

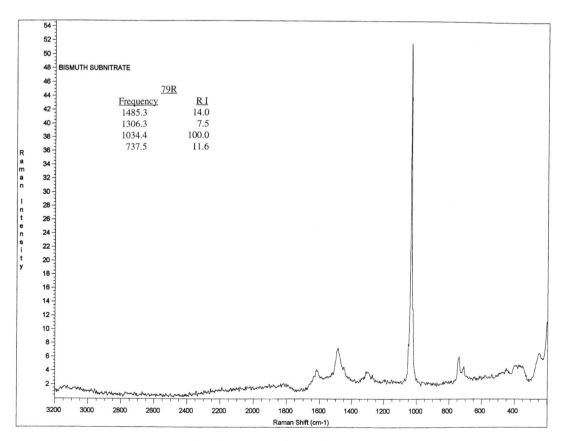

BISMUTH SUBNITRATE

79R

Frequency	R I
1485.3	14.0
1306.3	7.5
1034.4	100.0
737.5	11.6

79 Bismuth subnitrate $BiONO_3 \cdot H_2O$

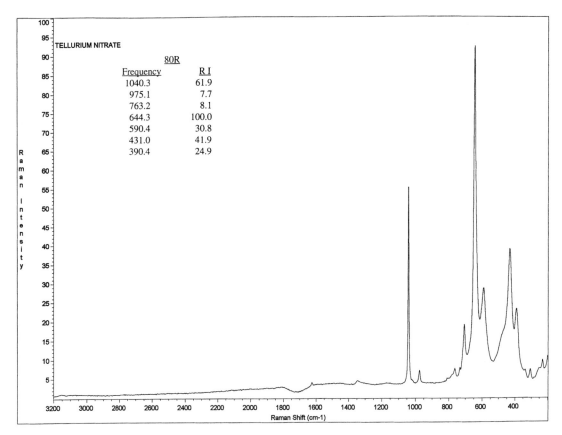

80   Tellurium nitrate (basic) $4TeO_2 \cdot N_2O_5 \cdot 1\frac{1}{2} H_2O$

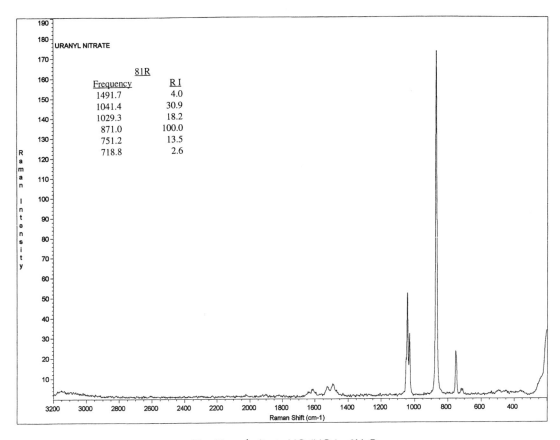

81   Uranyl nitrate $UO_2(NO_3)_2 \cdot 6H_2O$

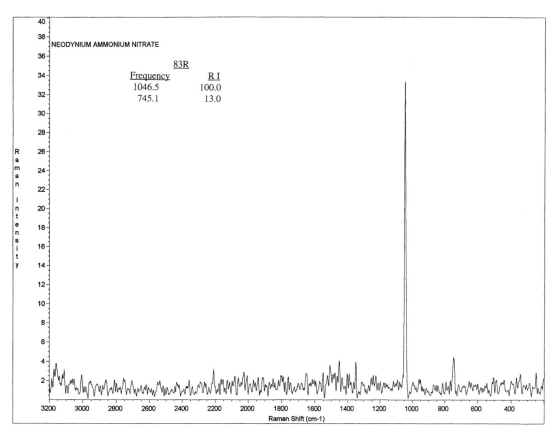

NEODYNIUM AMMONIUM NITRATE

83R

Frequency	R I
1046.5	100.0
745.1	13.0

83   Neodymium ammonium nitrate $NdNH_4(NO_3)_4$

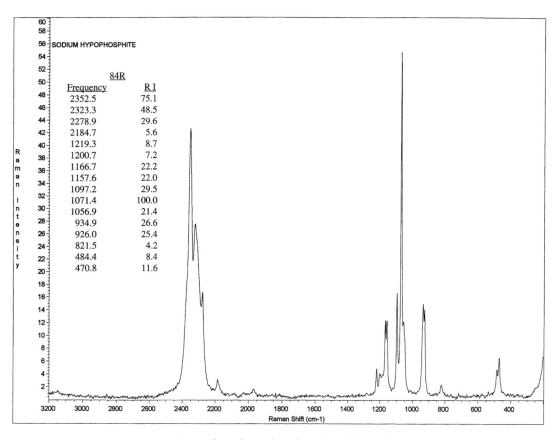

SODIUM HYPOPHOSPHITE

84R

Frequency	R I
2352.5	75.1
2323.3	48.5
2278.9	29.6
2184.7	5.6
1219.3	8.7
1200.7	7.2
1166.7	22.2
1157.6	22.0
1097.2	29.5
1071.4	100.0
1056.9	21.4
934.9	26.6
926.0	25.4
821.5	4.2
484.4	8.4
470.8	11.6

84   Sodium hypophosphite $NaH_2PO_2 \cdot H_2O$

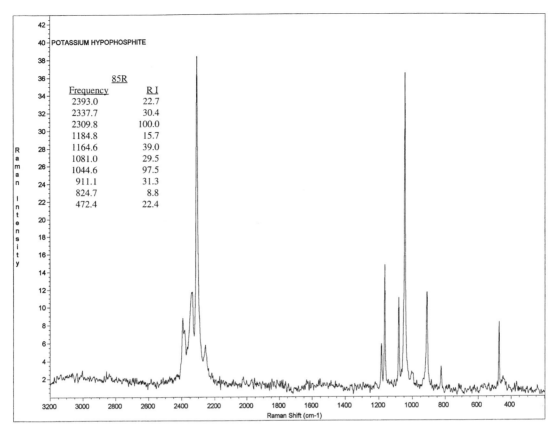

Frequency table (85R):

Frequency	R I
2393.0	22.7
2337.7	30.4
2309.8	100.0
1184.8	15.7
1164.6	39.0
1081.0	29.5
1044.6	97.5
911.1	31.3
824.7	8.8
472.4	22.4

85   Potassium hypophosphite $KH_2PO_2 \cdot xH_2O$

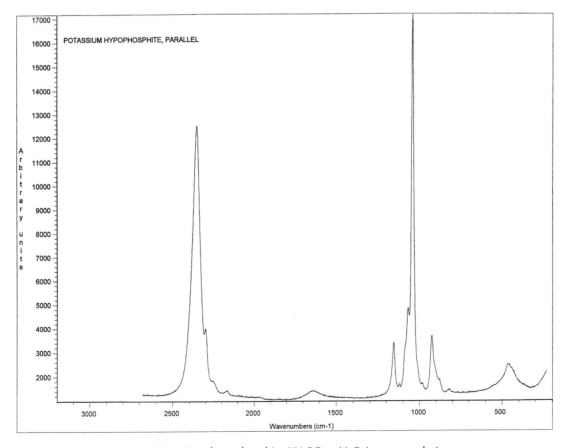

85a   Potassium hypophosphite $KH_2PO_2 \cdot xH_2O$ in water solution

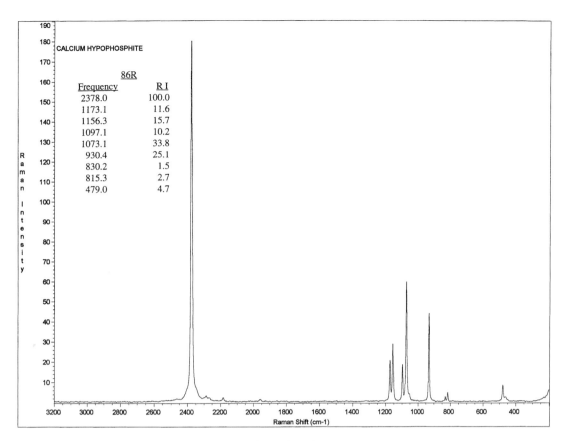

CALCIUM HYPOPHOSPHITE

86R

Frequency	R I
2378.0	100.0
1173.1	11.6
1156.3	15.7
1097.1	10.2
1073.1	33.8
930.4	25.1
830.2	1.5
815.3	2.7
479.0	4.7

86   Calcium hypophosphite $Ca(H_2PO_2)_2$

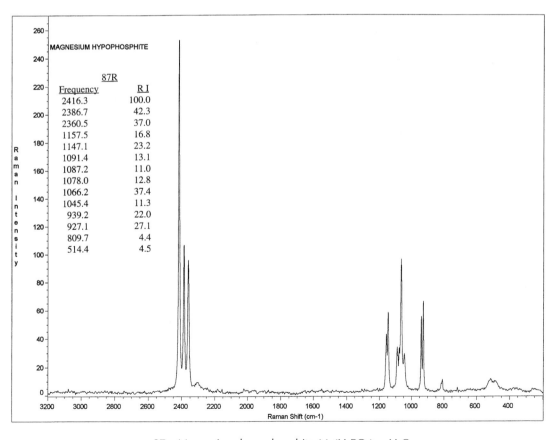

MAGNESIUM HYPOPHOSPHITE

87R

Frequency	R I
2416.3	100.0
2386.7	42.3
2360.5	37.0
1157.5	16.8
1147.1	23.2
1091.4	13.1
1087.2	11.0
1078.0	12.8
1066.2	37.4
1045.4	11.3
939.2	22.0
927.1	27.1
809.7	4.4
514.4	4.5

87   Magnesium hypophosphite $Mg(H_2PO_2)_2 \cdot xH_2O$

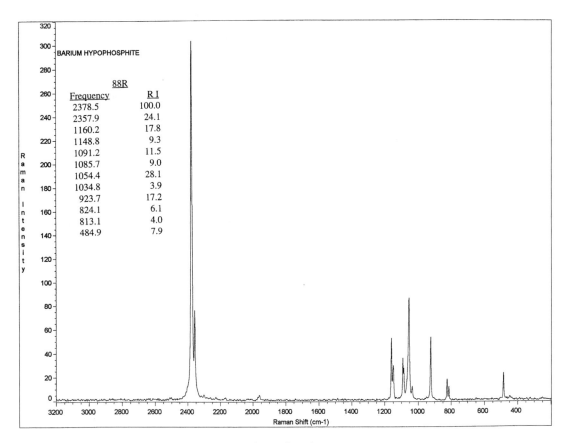

BARIUM HYPOPHOSPHITE

88R	
Frequency	R I
2378.5	100.0
2357.9	24.1
1160.2	17.8
1148.8	9.3
1091.2	11.5
1085.7	9.0
1054.4	28.1
1034.8	3.9
923.7	17.2
824.1	6.1
813.1	4.0
484.9	7.9

88   Barium hypophosphite $Ba(H_2PO_2)_2$

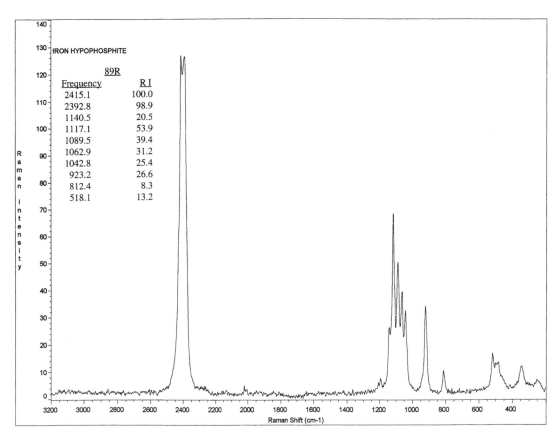

IRON HYPOPHOSPHITE

89R	
Frequency	R I
2415.1	100.0
2392.8	98.9
1140.5	20.5
1117.1	53.9
1089.5	39.4
1062.9	31.2
1042.8	25.4
923.2	26.6
812.4	8.3
518.1	13.2

89   Iron (III) hypophosphite $Fe(H_2PO_2)_3$

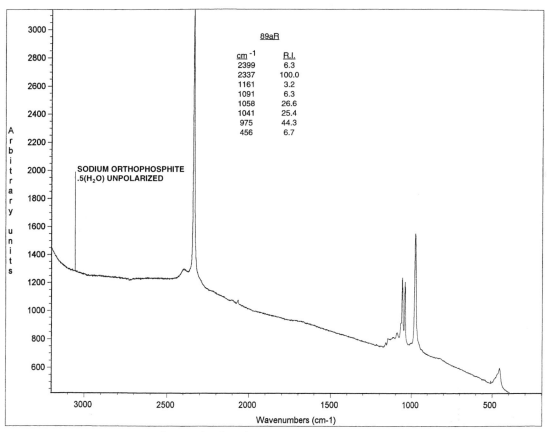

89a   Sodium orthophosphite Na$_2$HPO$_3$·5H$_2$O in water solution

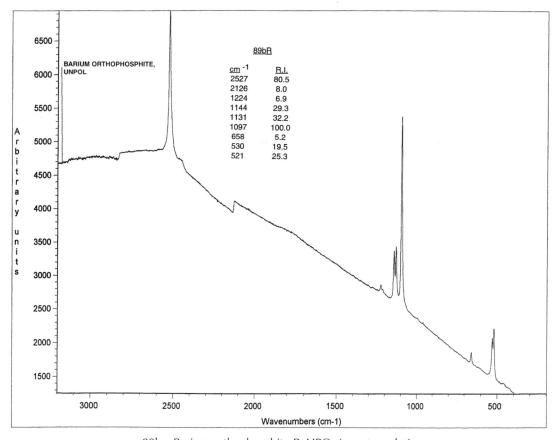

89b   Barium orthophosphite BaHPO$_3$ in water solution

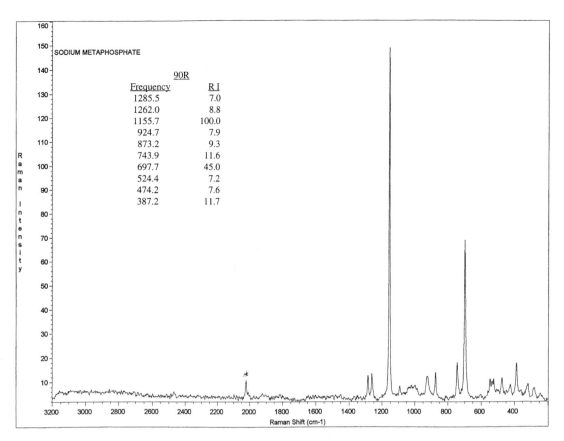

90   Sodium metaphosphate $(NaPO_3)_x \cdot xH_2O$

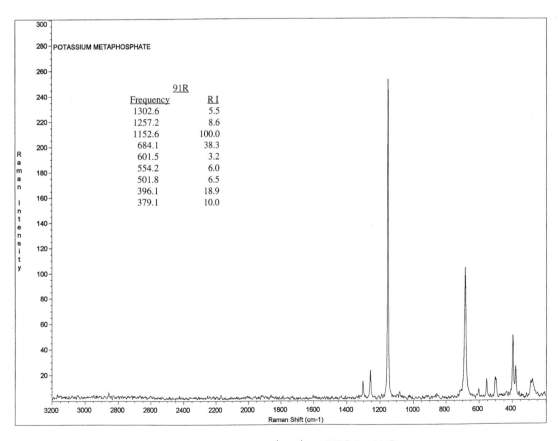

91   Potassium metaphosphate $(KPO_3)_x \cdot xH_2O$

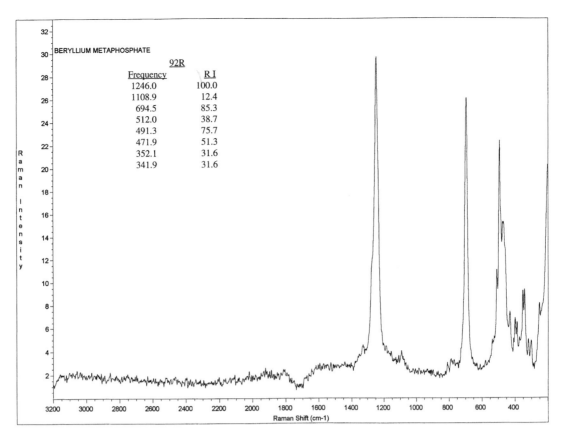

92R	
Frequency	R I
1246.0	100.0
1108.9	12.4
694.5	85.3
512.0	38.7
491.3	75.7
471.9	51.3
352.1	31.6
341.9	31.6

92   Beryllium metaphosphate $[Be(PO_3)_2]_x \cdot xH_2O$

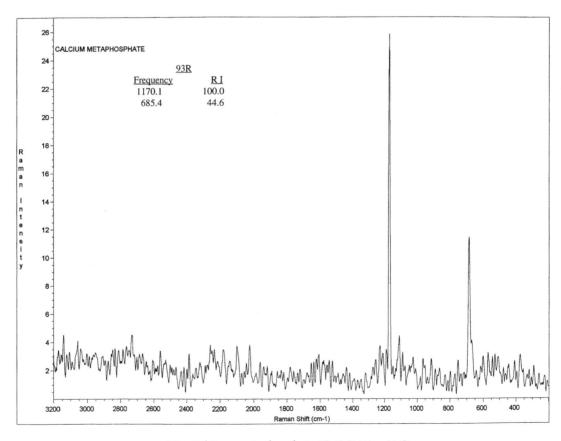

93R	
Frequency	R I
1170.1	100.0
685.4	44.6

93   Calcium metaphosphate $[Ca(PO_3)_2)]_x \cdot xH_2O$

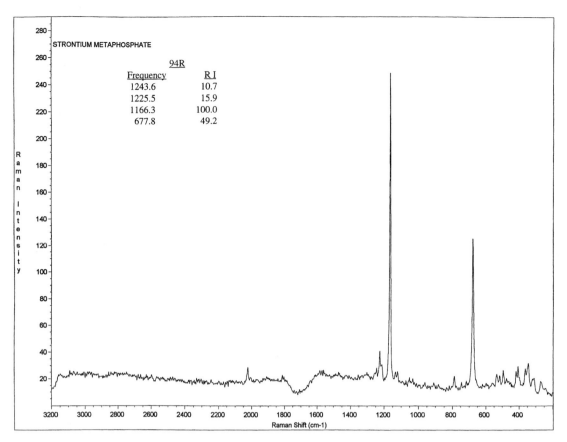

94 Strontium metaphosphate [Sr(PO$_3$)$_2$]$_x$·xH$_2$O

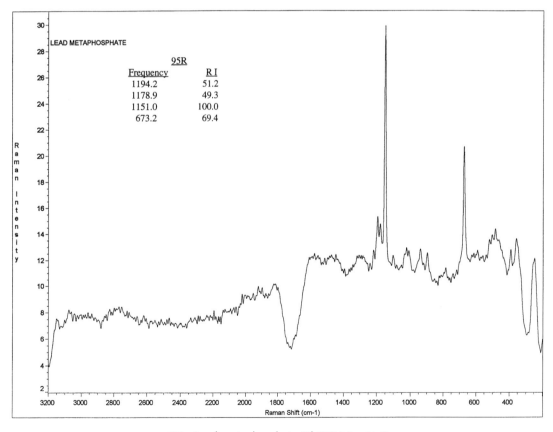

95 Lead metaphosphate [Pb(PO$_3$)$_2$]$_x$·xH$_2$O

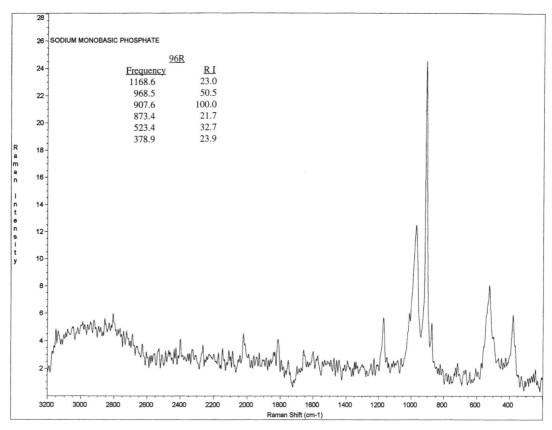

SODIUM MONOBASIC PHOSPHATE

96R

Frequency	R I
1168.6	23.0
968.5	50.5
907.6	100.0
873.4	21.7
523.4	32.7
378.9	23.9

96   Sodium orthophoshate (monobasic) $NaH_2PO_4 \cdot xH_2O$

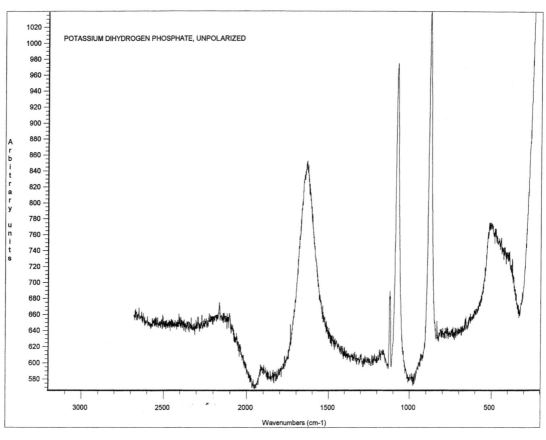

POTASSIUM DIHYDROGEN PHOSPHATE, UNPOLARIZED

96a   Potassium orthophosphate (monobasic) KH_2PO_4 in water solution

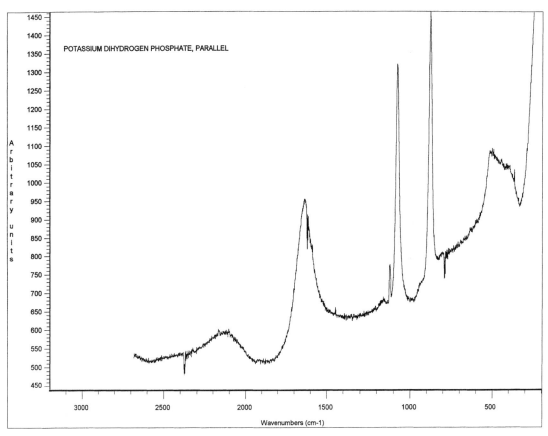

96b   Potassium orthophosphate (monobasic) KH$_2$PO$_4$ in water solution

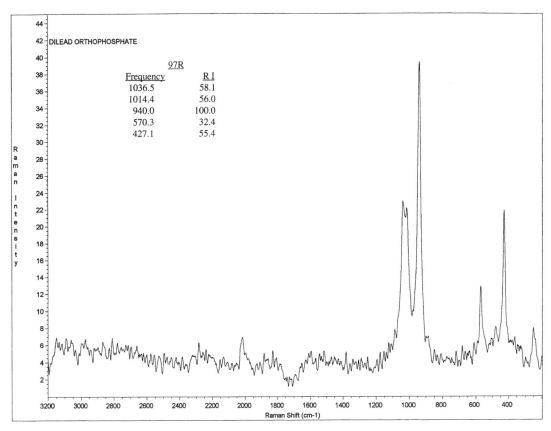

97   Lead orthophosphate (monobasic) Pb(H$_2$PO$_4$)$_2$

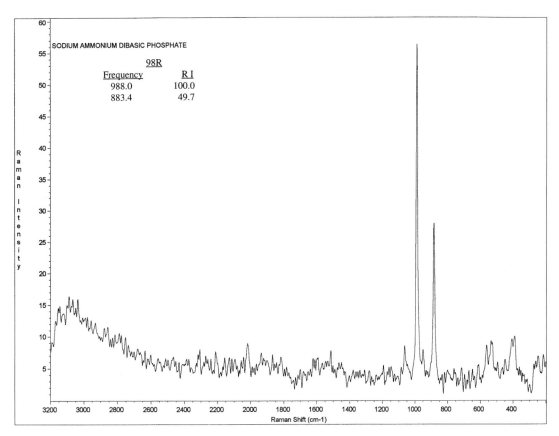

98   Sodium ammonium orthophosphate (dibasic) NaNH$_4$HPO$_4$

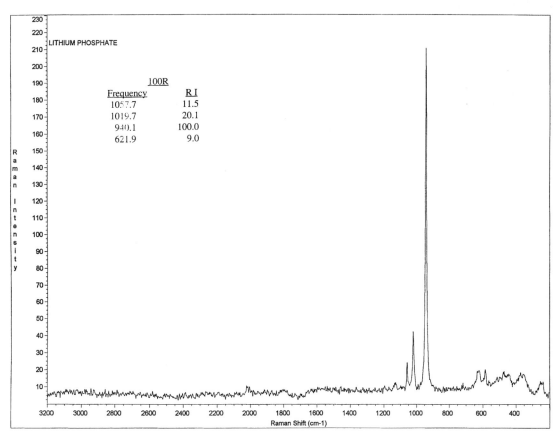

100   Lithium orthophosphate Li$_3$PO$_4$ · $^1/_2$H$_2$O

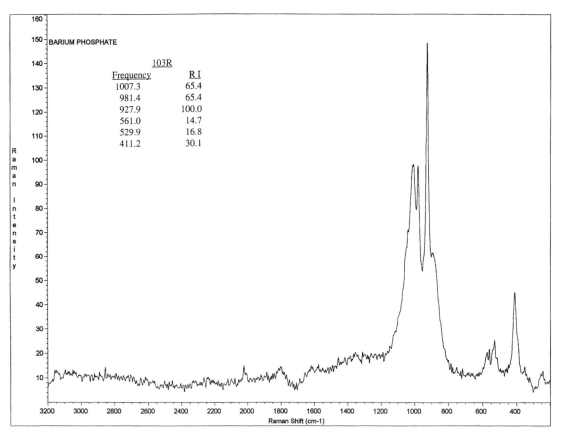

103   Barium orthophosphate Ba$_3$(PO$_4$)$_2$

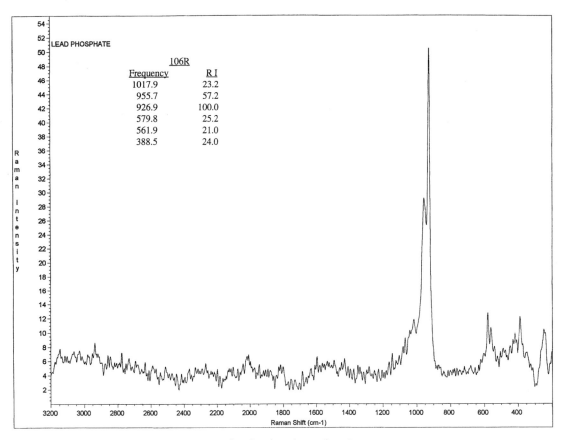

106   Lead orthophosphate Pb$_3$(PO$_4$)$_2$ wet

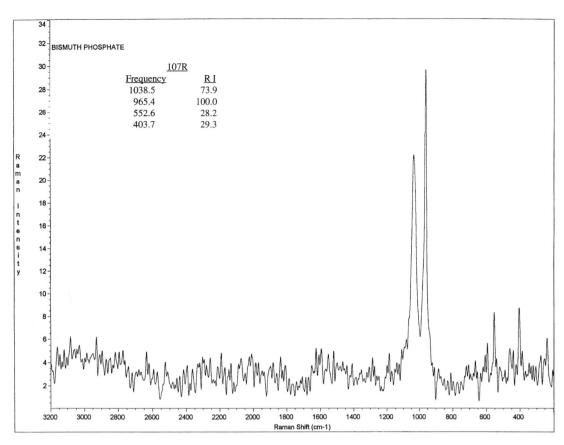

107   Bismuth orthophosphate BiPO$_4$

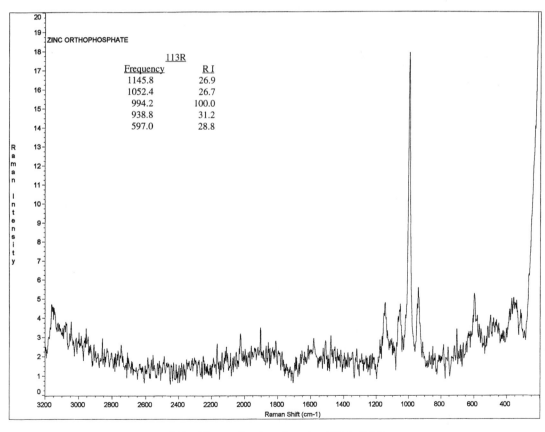

113   Zinc orthophosphate Zn$_3$(PO$_4$)·4H$_2$O

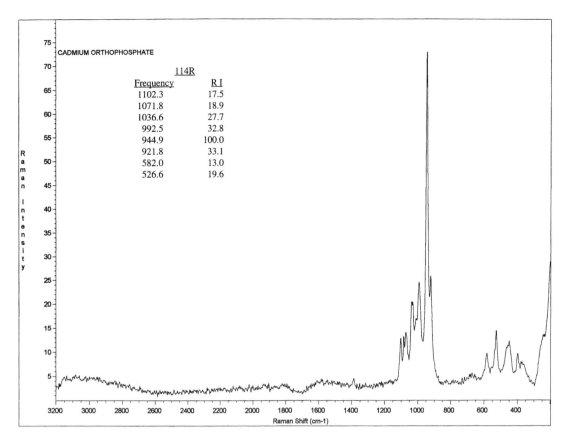

CADMIUM ORTHOPHOSPHATE

114R

Frequency	R I
1102.3	17.5
1071.8	18.9
1036.6	27.7
992.5	32.8
944.9	100.0
921.8	33.1
582.0	13.0
526.6	19.6

114   Cadmium orthophosphate $Cd_3(PO_4) \cdot xH_2O$

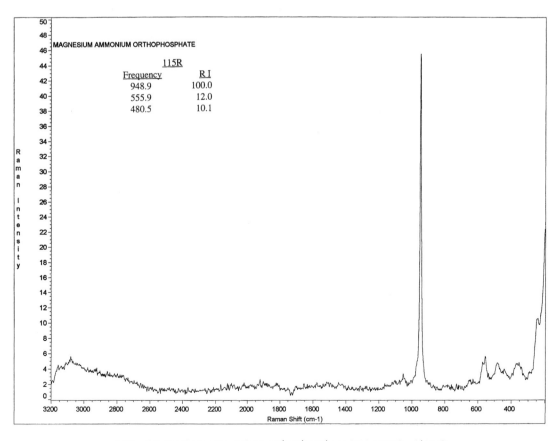

MAGNESIUM AMMONIUM ORTHOPHOSPHATE

115R

Frequency	R I
948.9	100.0
555.9	12.0
480.5	10.1

115   Magnesium ammonium orthophosphate $NH_4MgPO_4 \cdot \frac{1}{2}H_2O$

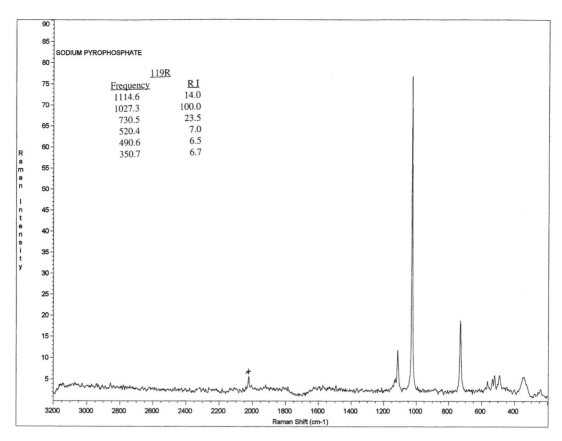

119   Sodium pyrophosphate Na$_4$P$_2$O$_7$

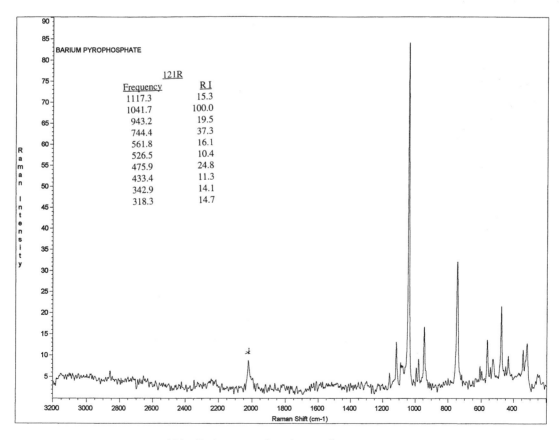

121   Barium pyrophosphate (α form) Ba$_2$P$_2$O$_7$

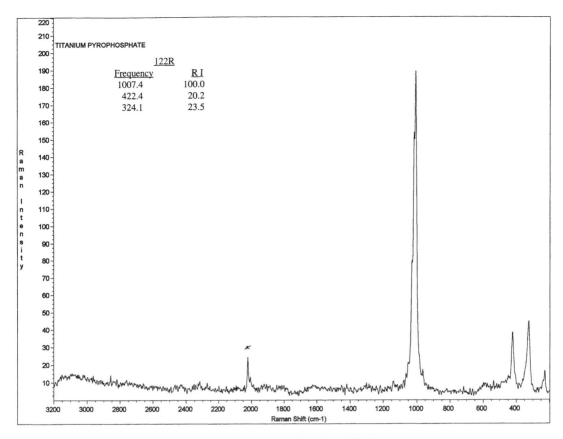

122 Titanium pyrophosphate TiP$_2$O$_4$

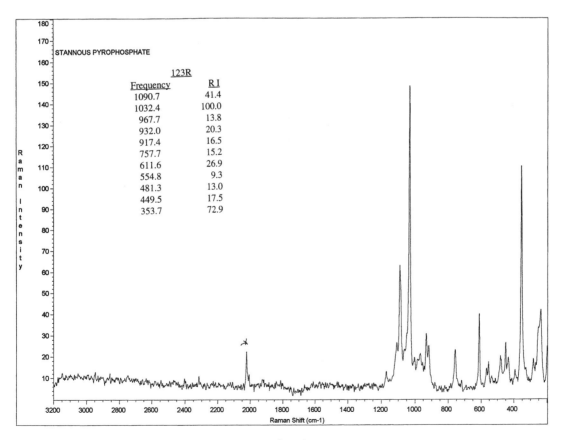

123 Tin pyrophosphate Sn$_2$P$_2$O$_7$

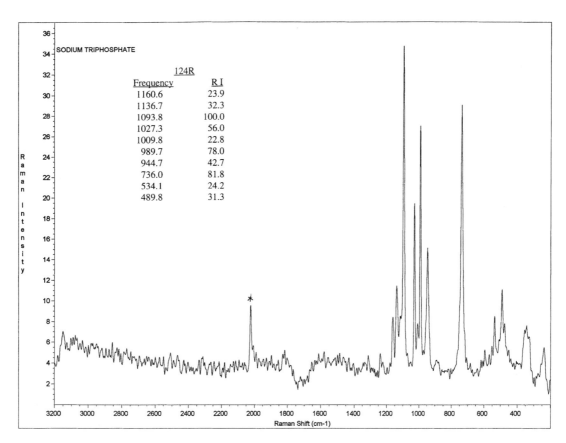

124   Sodium tripolyphosphate Na$_5$P$_3$O$_{10}$·xH$_2$O

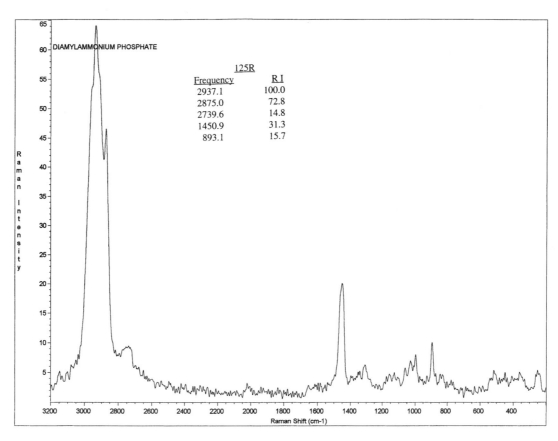

125   Diamylammonium phosphate (C$_5$H$_{11}$O)$_2$PO$_2$NH$_4$

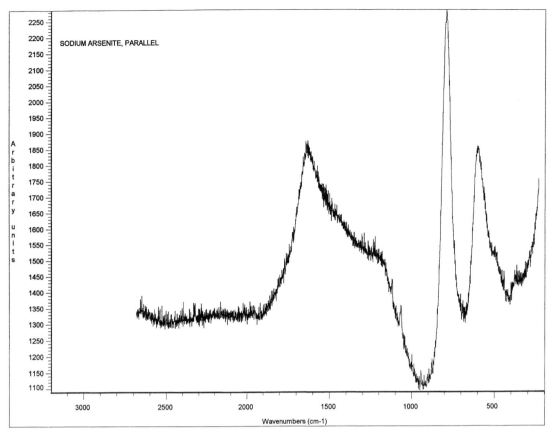

125a   Sodium metaarsenite Na_2AsO_2 in water solution

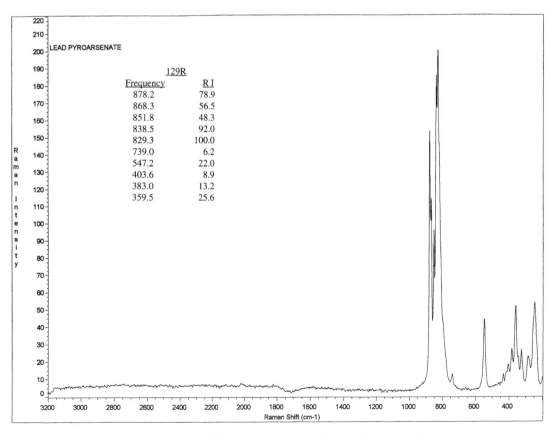

129   Lead pyroarsenate (monobasic) $Pb_2As_2O_7$

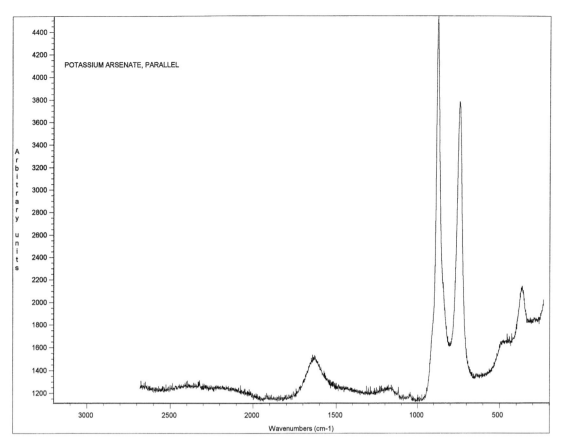

129a   Potassium orthoarsenate (monobasic) KH$_2$AsO$_4$ in water solution

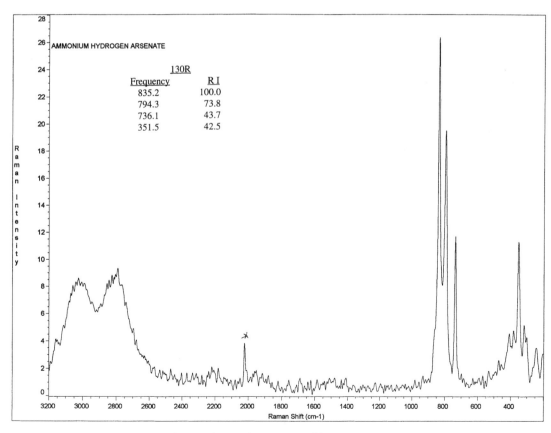

130   Ammonium orthoarsenate (dibasic) (NH$_4$)$_2$HAsO$_4$

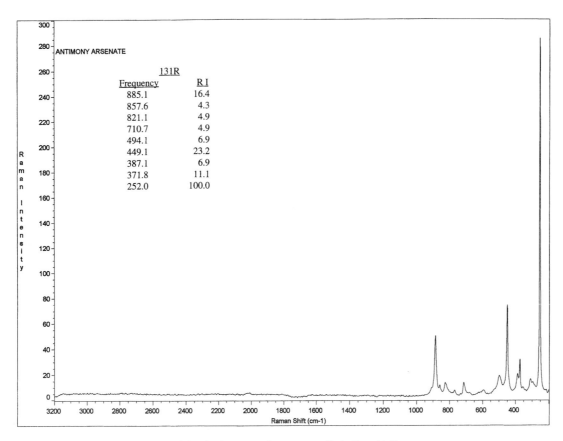

131   Antimony orthoarsenate SbAsO$_4$·xH$_2$O

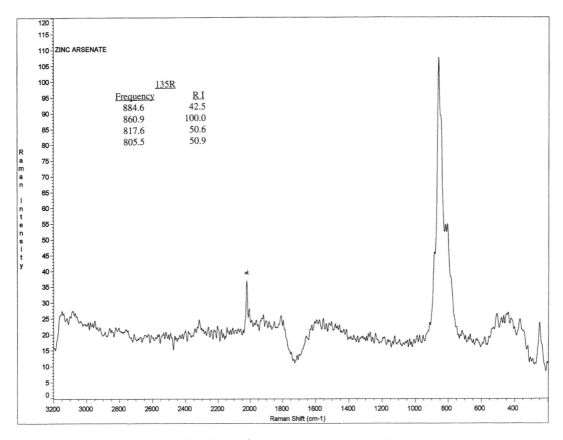

135   Zinc orthoarsenate Zn$_3$(AsO$_4$)$_2$·8H$_2$O

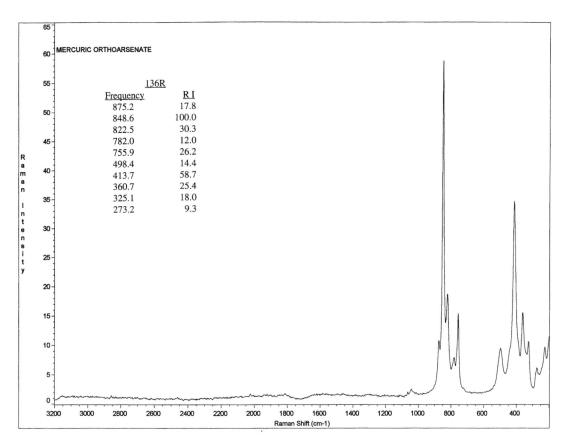

136R	
Frequency	R I
875.2	17.8
848.6	100.0
822.5	30.3
782.0	12.0
755.9	26.2
498.4	14.4
413.7	58.7
360.7	25.4
325.1	18.0
273.2	9.3

136   Mercury (II) orthoarsenate $Hg_3(AsO_4)_2$

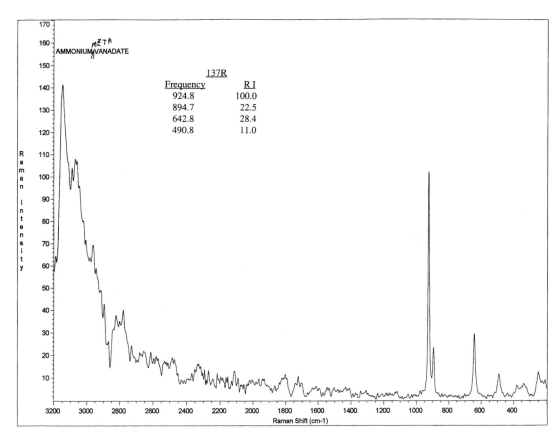

137R	
Frequency	R I
924.8	100.0
894.7	22.5
642.8	28.4
490.8	11.0

137   Ammonium metavanadate NH_4VO_3

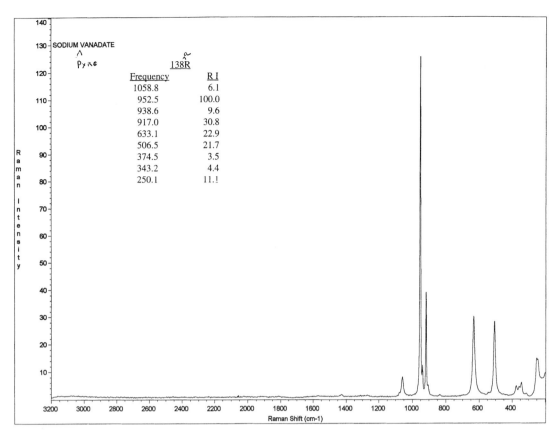

138a   Sodium pyrovanadate $Na_4V_2O_7 \cdot xH_2O$

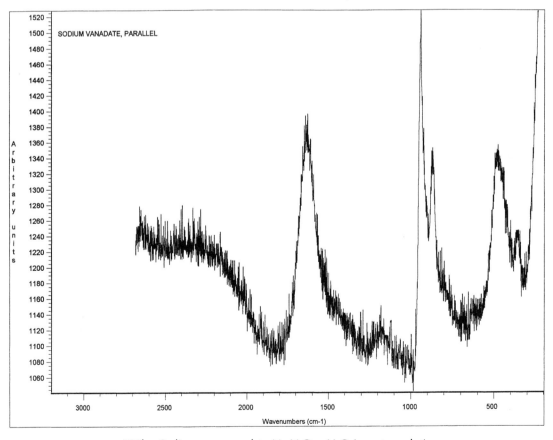

138b   Sodium pyrovanadate $Na_4V_2O_7 \cdot xH_2O$ in water solution

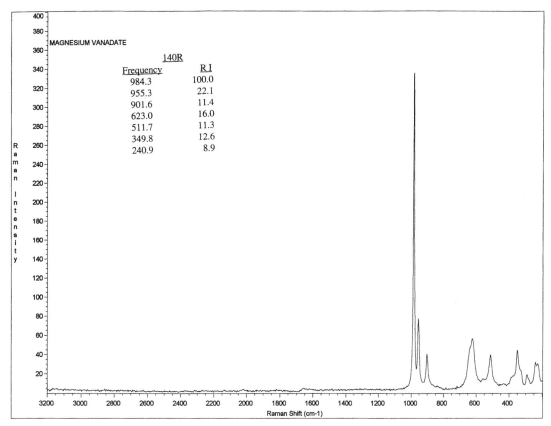

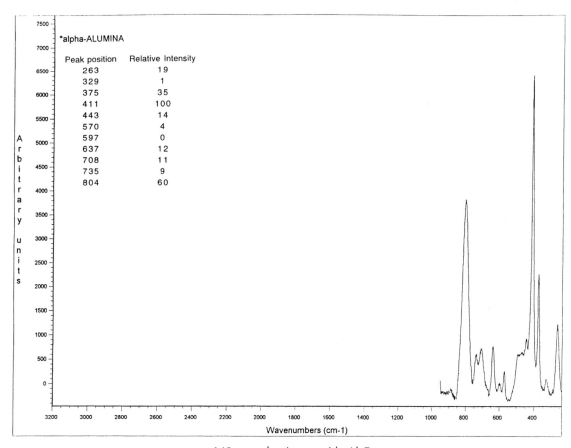

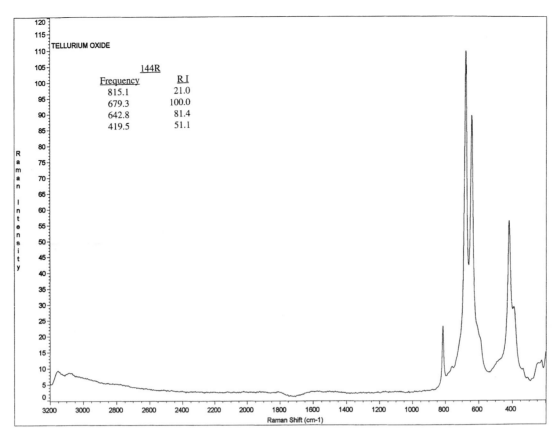

144   Tellurium dioxide TeO$_2$

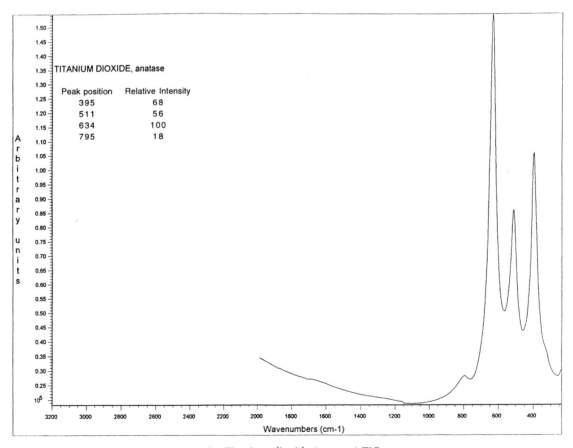

145   Titanium dioxide (anatase) TiO$_2$

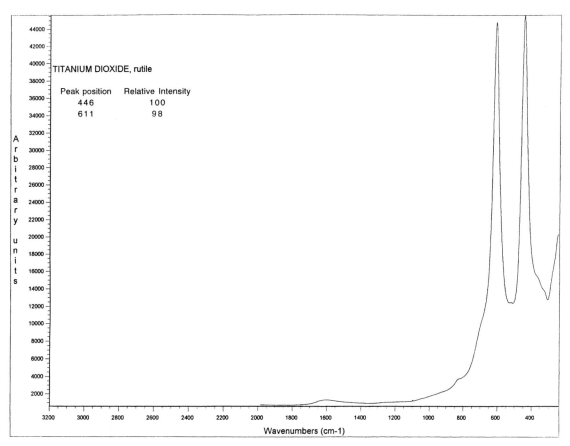

146   Titanium dioxide (rutile) TiO$_2$

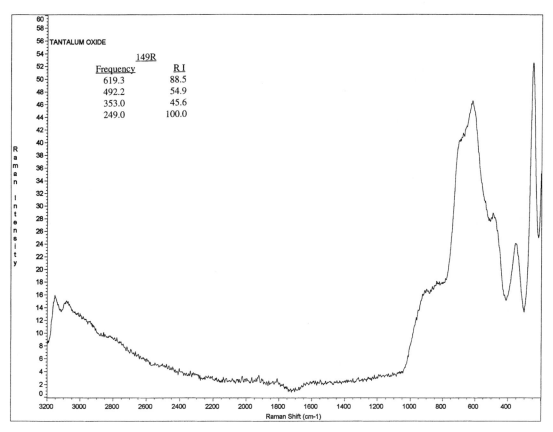

149   Tantalum pentoxide Ta$_2$O$_5$ wet

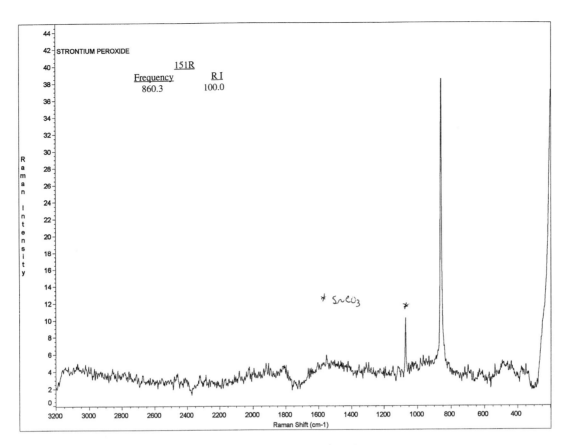

151   Strontium peroxide SrO₂

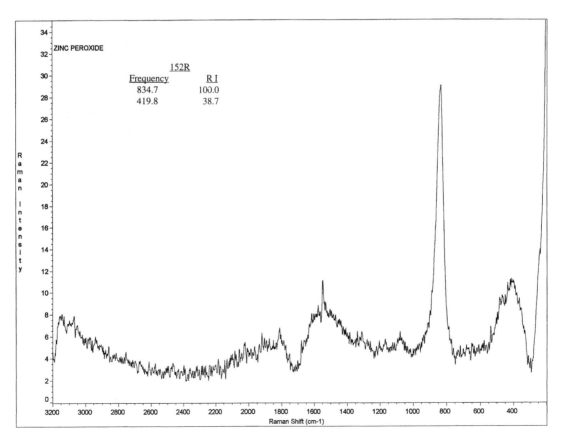

152   Zinc peroxide ZnO₂

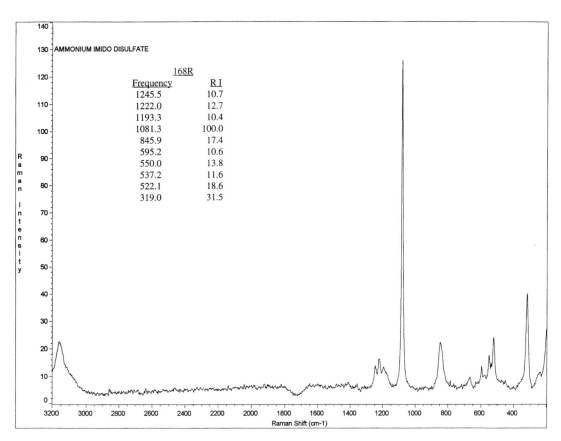

AMMONIUM IMIDO DISULFATE

168R

Frequency	R I
1245.5	10.7
1222.0	12.7
1193.3	10.4
1081.3	100.0
845.9	17.4
595.2	10.6
550.0	13.8
537.2	11.6
522.1	18.6
319.0	31.5

168   Ammonium imidodisulfate $(NH_4)_2S_2NHO_6$

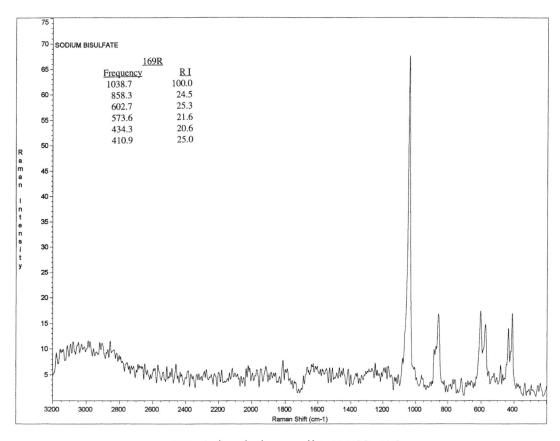

SODIUM BISULFATE

169R

Frequency	R I
1038.7	100.0
858.3	24.5
602.7	25.3
573.6	21.6
434.3	20.6
410.9	25.0

169   Sodium hydrogen sulfate $NaHSO_4 \cdot H_2O$

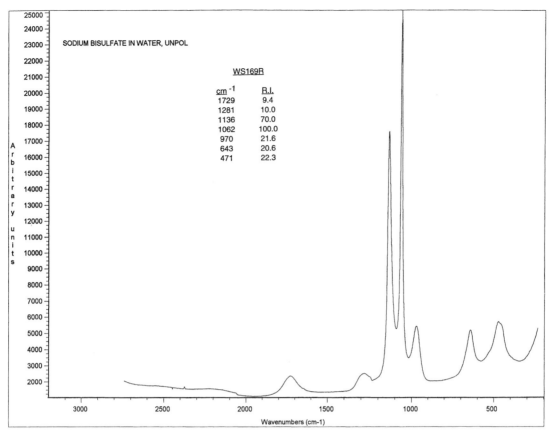

169a   Sodium hydrogen sulfate NaHSO$_4 \cdot$H$_2$O in water solution

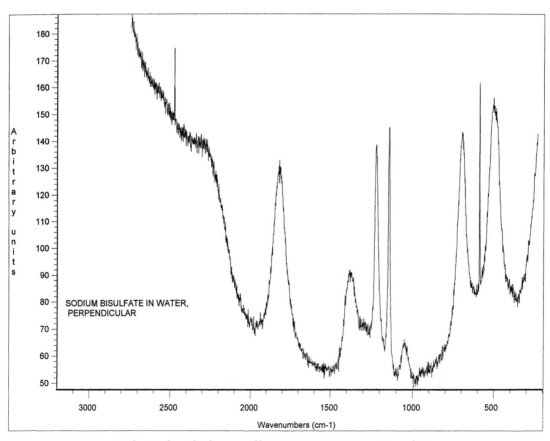

169b   Sodium hydrogen sulfate NaHSO$_4 \cdot$H$_2$O in water solution

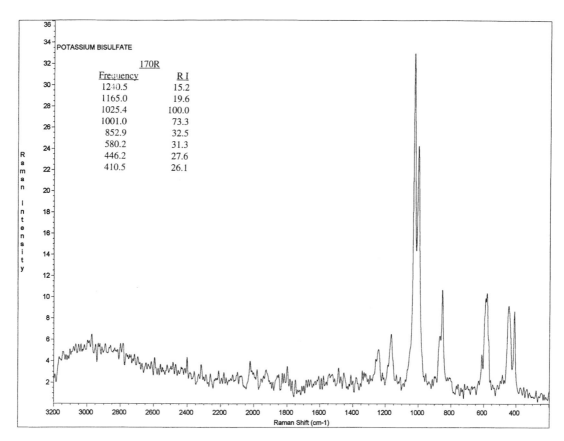

POTASSIUM BISULFATE

170R

Frequency	R I
1240.5	15.2
1165.0	19.6
1025.4	100.0
1001.0	73.3
852.9	32.5
580.2	31.3
446.2	27.6
410.5	26.1

170   Potassium hydrogen sulfate $KHSO_4$

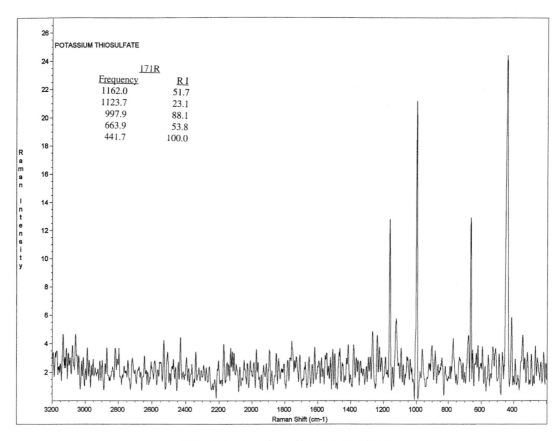

POTASSIUM THIOSULFATE

171R

Frequency	R I
1162.0	51.7
1123.7	23.1
997.9	88.1
663.9	53.8
441.7	100.0

171   Potassium thiosulfate $K_2S_2O_3 \cdot \frac{1}{3}H_2O$

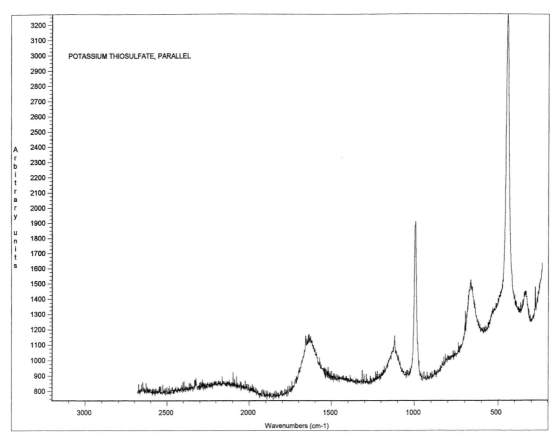

171a   Potassium thiosulfate $K_2S_2O_3 \cdot \frac{1}{3}H_2O$ in water solution

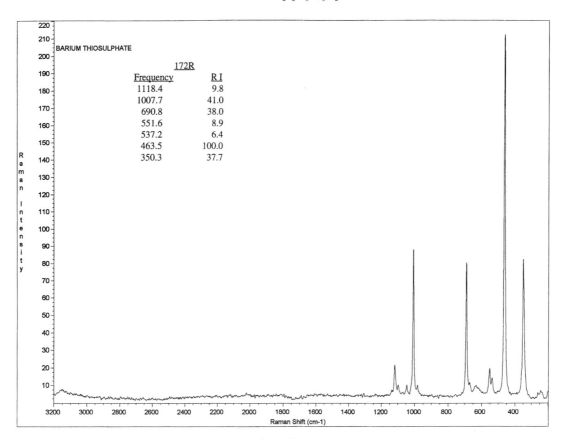

172R	
Frequency	R I
1118.4	9.8
1007.7	41.0
690.8	38.0
551.6	8.9
537.2	6.4
463.5	100.0
350.3	37.7

172   Barium thiosulfate $BaS_2O_3 \cdot H_2O$

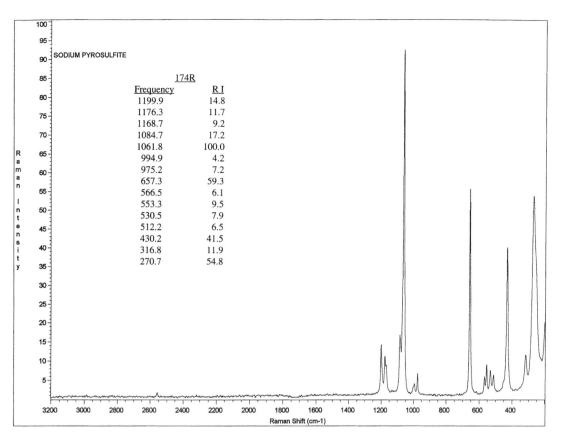

SODIUM PYROSULFITE

174R

Frequency	R I
1199.9	14.8
1176.3	11.7
1168.7	9.2
1084.7	17.2
1061.8	100.0
994.9	4.2
975.2	7.2
657.3	59.3
566.5	6.1
553.3	9.5
530.5	7.9
512.2	6.5
430.2	41.5
316.8	11.9
270.7	54.8

174   Sodium pyrosulfite $Na_2S_2O_5$

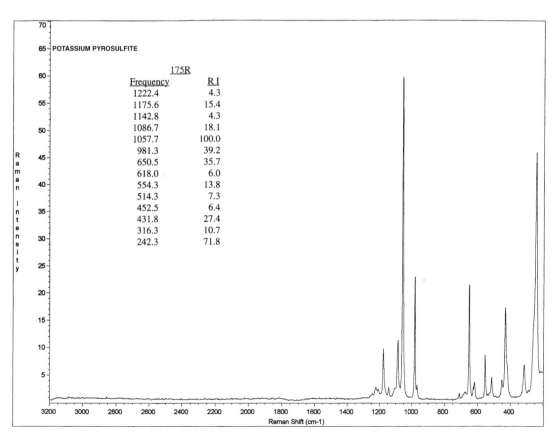

POTASSIUM PYROSULFITE

175R

Frequency	R I
1222.4	4.3
1175.6	15.4
1142.8	4.3
1086.7	18.1
1057.7	100.0
981.3	39.2
650.5	35.7
618.0	6.0
554.3	13.8
514.3	7.3
452.5	6.4
431.8	27.4
316.3	10.7
242.3	71.8

175   Potassium pyrosulfite $K_2S_2O_5$

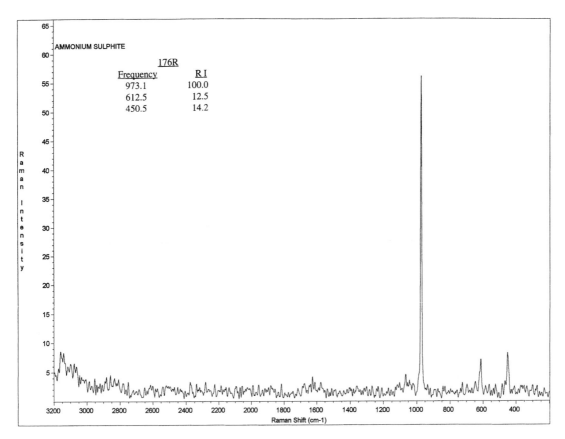

176   Ammonium sulfite $(NH_4)_2SO_3$

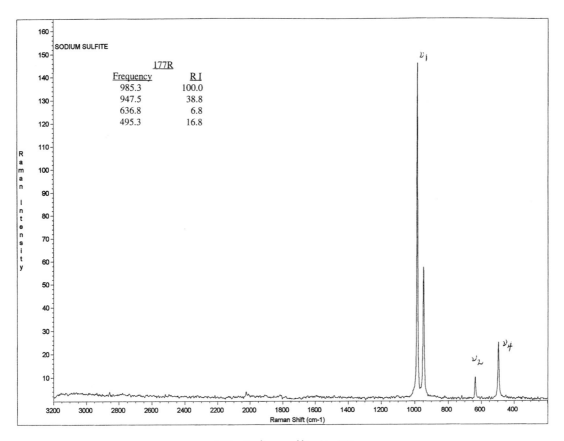

177   Sodium sulfite Na_2SO_3

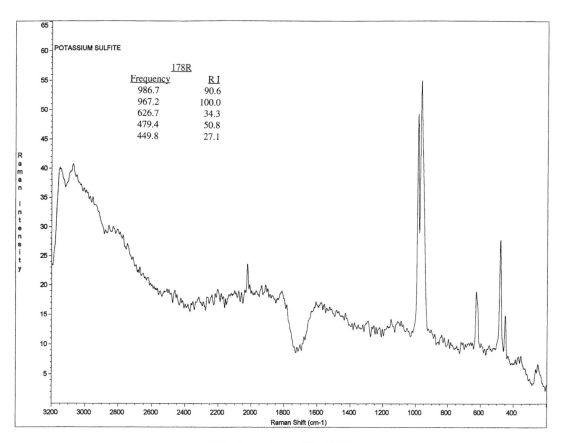

178   Potassium sulfite K$_2$SO$_3$

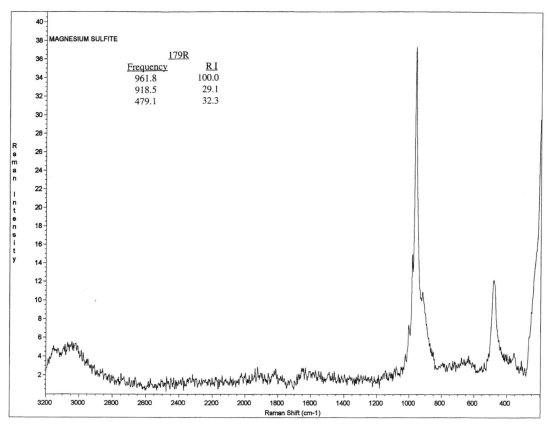

179   Magnesium sulfite MgSO$_3 \cdot$ xH$_2$O

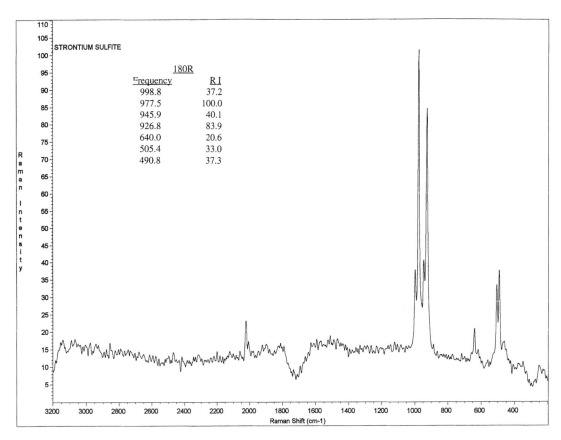

180   Strontium sulfite SrSO₃

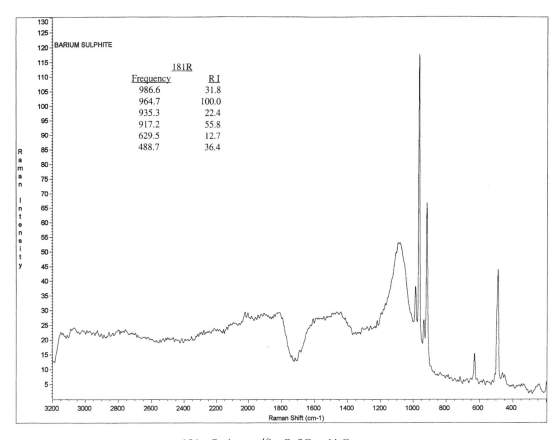

181   Barium sulfite BaSO₃·xH₂O or wet

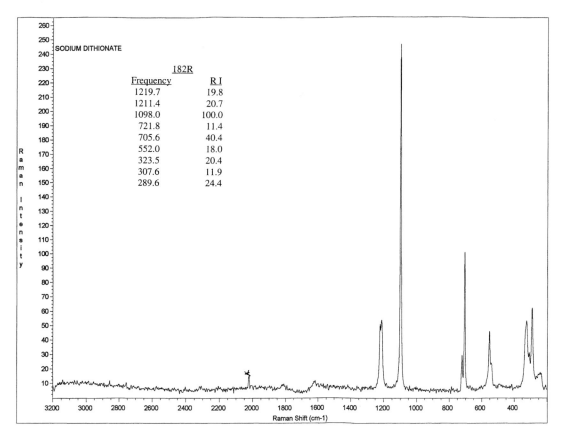

182   Sodium dithionate $Na_2S_2O_6$

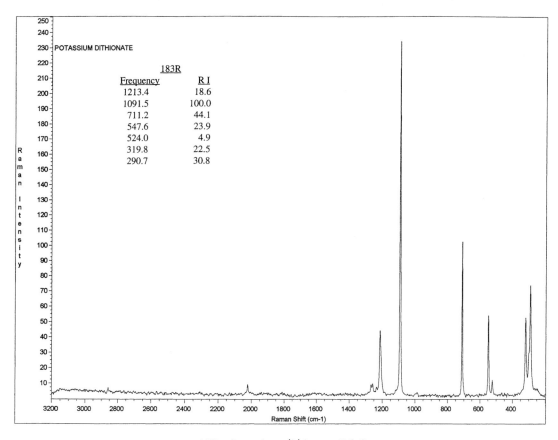

183   Potassium dithionate $K_2S_2O_6$

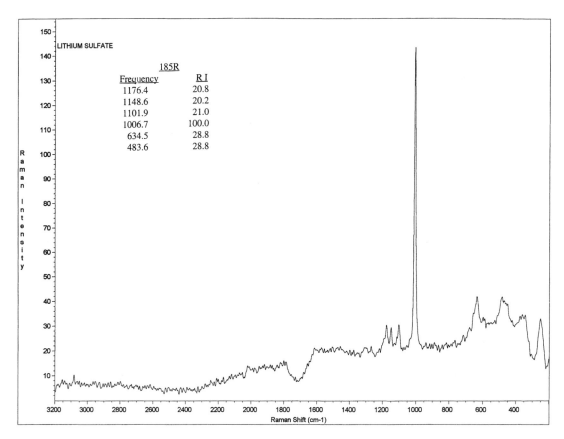

185   Lithium sulfate Li$_2$SO$_4 \cdot$H$_2$O

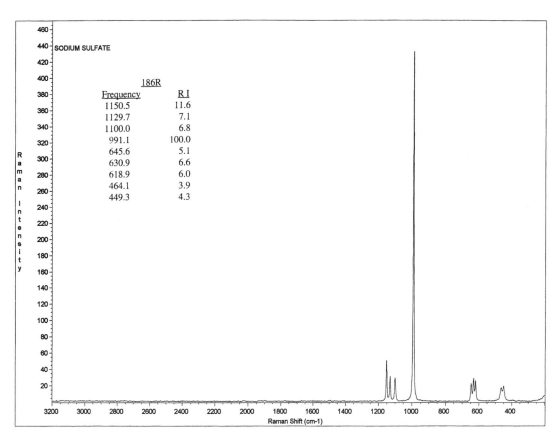

186   Sodium sulfate Na$_2$SO$_4$

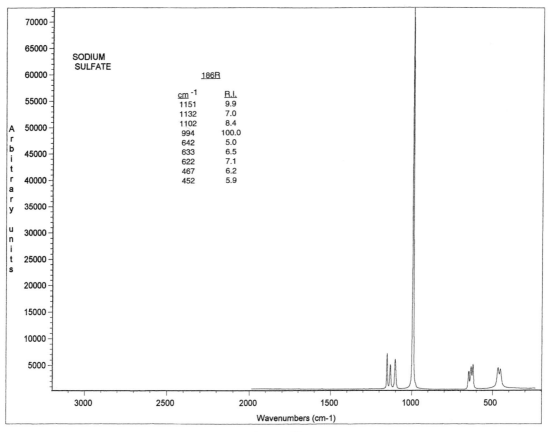

SODIUM
SULFATE

186R

cm^{-1}	R.I.
1151	9.9
1132	7.0
1102	8.4
994	100.0
642	5.0
633	6.5
622	7.1
467	6.2
452	5.9

186a   Sodium sulfate Na$_2$SO$_4$

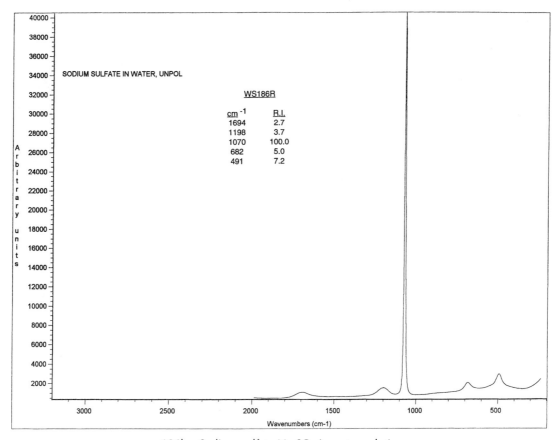

SODIUM SULFATE IN WATER, UNPOL

WS186R

cm^{-1}	R.I.
1694	2.7
1198	3.7
1070	100.0
682	5.0
491	7.2

186b   Sodium sulfate Na$_2$SO$_4$ in water solution

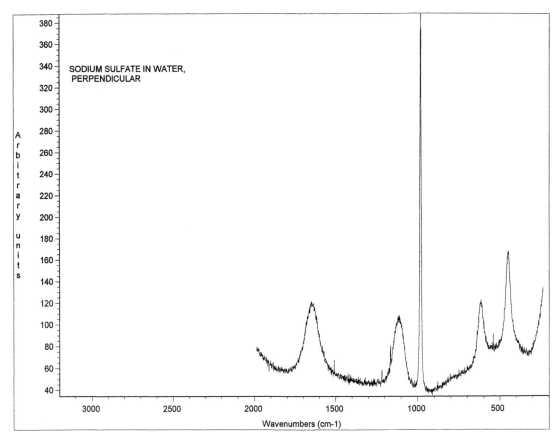

186c   Sodium sulfate Na$_2$SO$_4$ in water solution

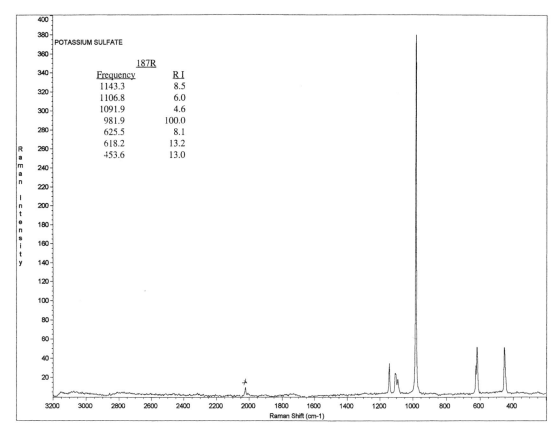

POTASSIUM SULFATE

187R	
Frequency	R I
1143.3	8.5
1106.8	6.0
1091.9	4.6
981.9	100.0
625.5	8.1
618.2	13.2
453.6	13.0

187   Potassium sulfate K$_2$SO$_4$

116

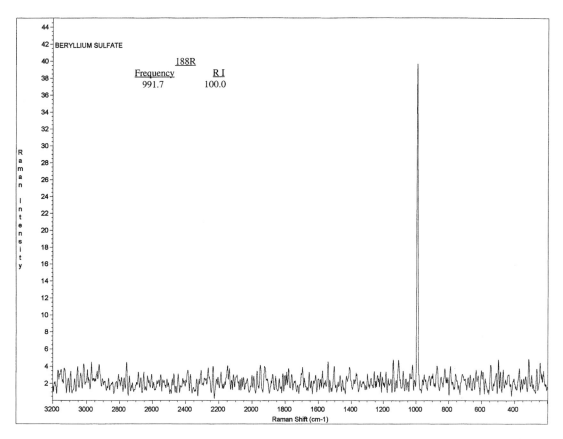

188   Beryllium sulfate $BeSO_4 \cdot 4H_2O$

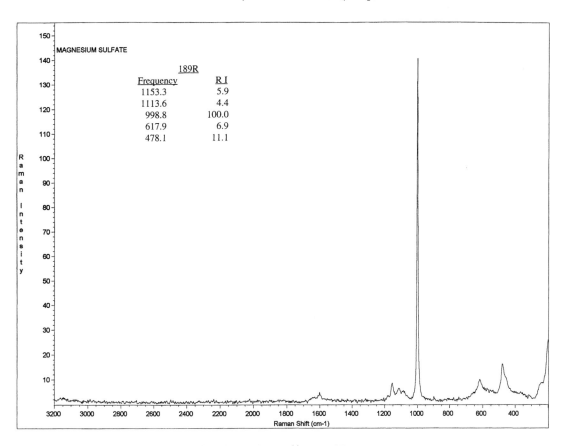

189   Magnesium sulfate $MgSO_4 \cdot 7H_2O$

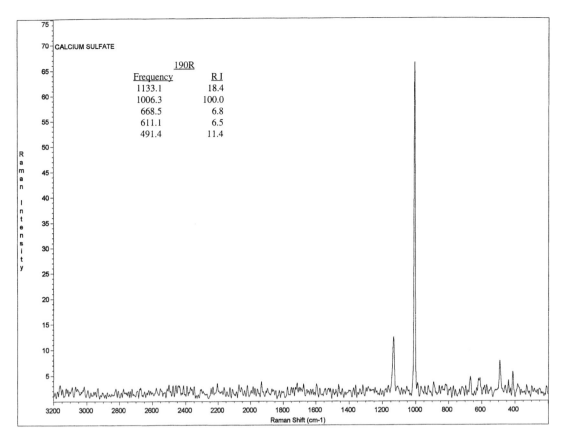

190   Calcium sulfate CaSO$_4 \cdot 2$H$_2$O

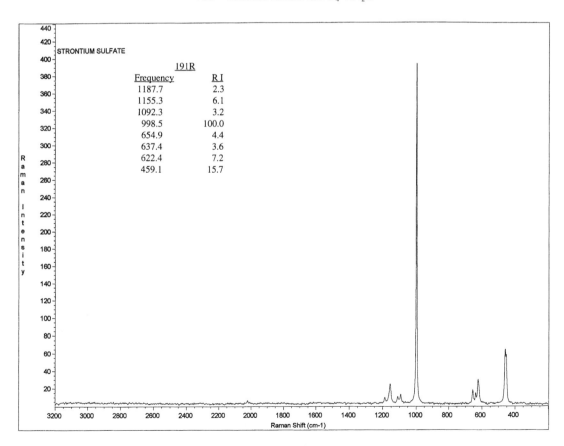

191   Strontium sulfate SrSO$_4$

118

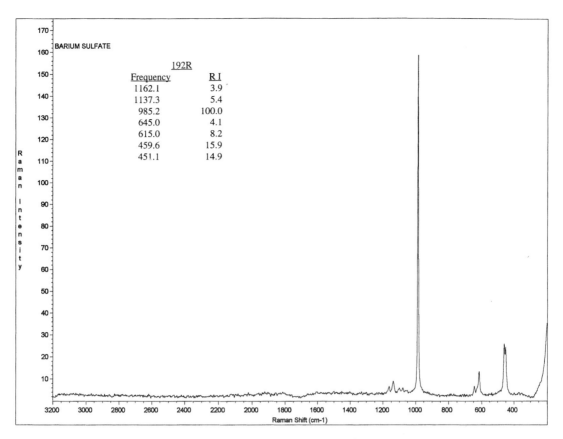

192   Barium sulfate BaSO$_4$

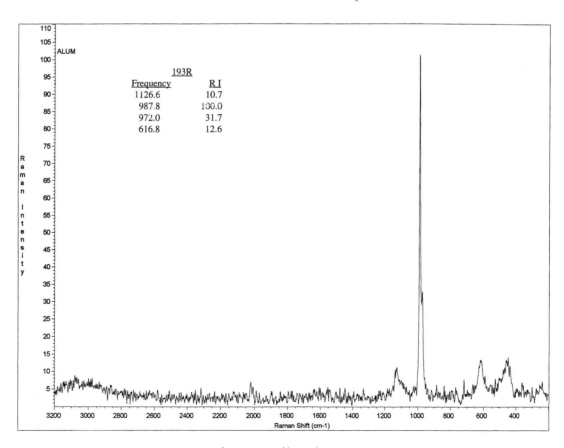

193   Aluminum sulfate Al$_2$(SO$_4$)$_3$·18H$_2$O

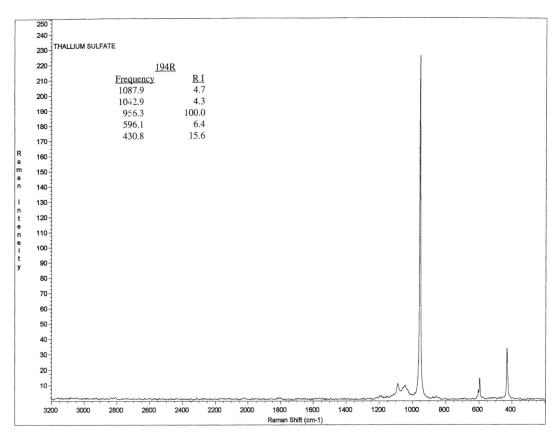

THALLIUM SULFATE

194R

Frequency	R I
1087.9	4.7
1042.9	4.3
956.3	100.0
596.1	6.4
430.8	15.6

194   Thallium sulfate Tl(SO$_4$)$_3$

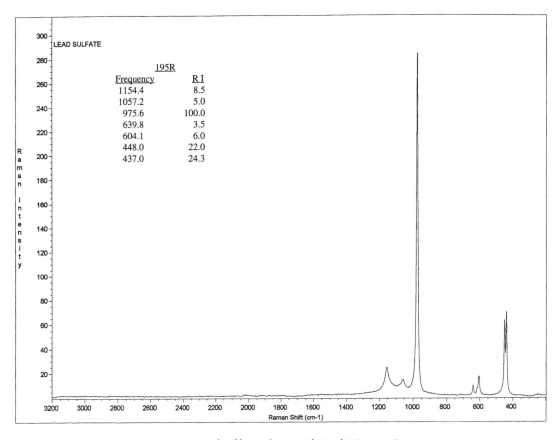

LEAD SULFATE

195R

Frequency	R I
1154.4	8.5
1057.2	5.0
975.6	100.0
639.8	3.5
604.1	6.0
448.0	22.0
437.0	24.3

195   Lead sulfate tribasic 3PbO·PbSO$_4$·xH$_2$O

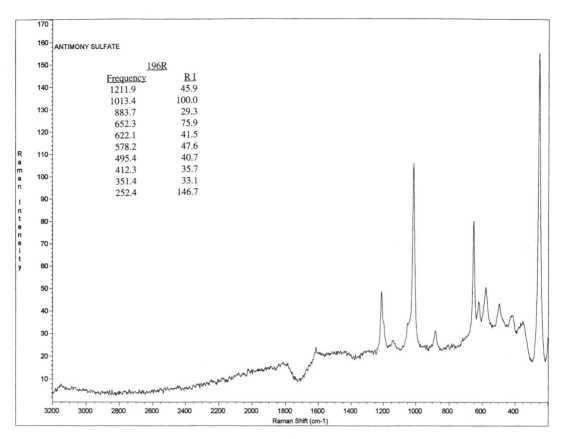

196   Antimony sulfate $Sb_2(SO_4)_3 \cdot xH_2O$

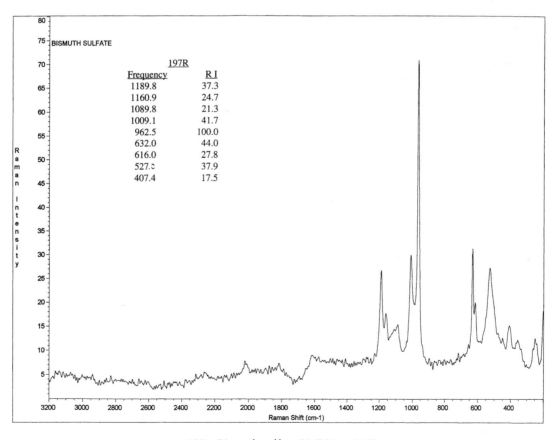

197   Bismuth sulfate $Bi_2(SO_4)_3 \cdot xH_2O$

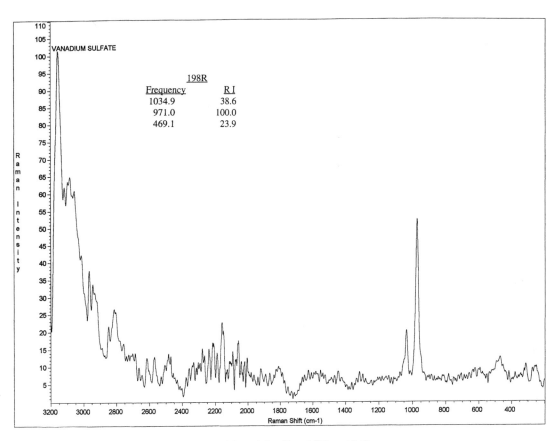

198   Vanadyl sulfate VSO$_4 \cdot$xH$_2$O

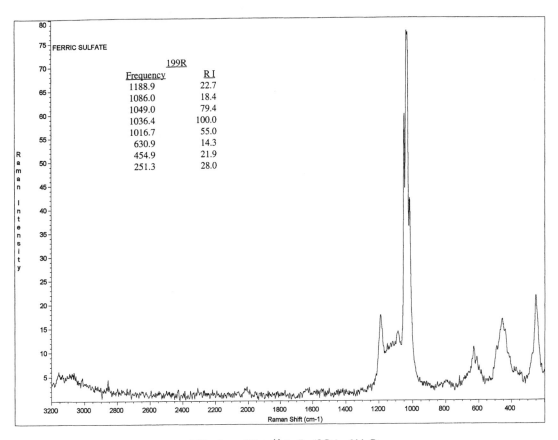

199   Iron (III) sulfate Fe$_2$(SO$_4$)$_3 \cdot$9H$_2$O

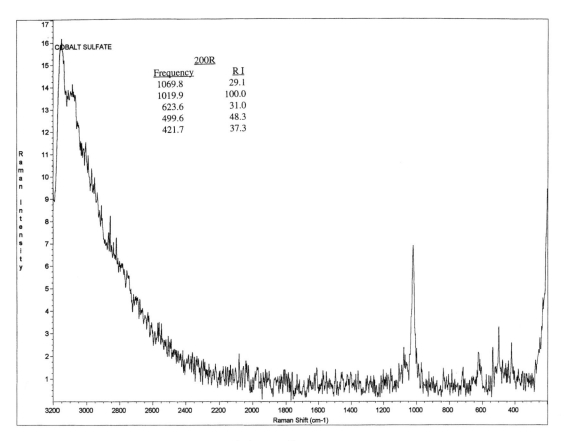

	200R	
Frequency		R I
1069.8		29.1
1019.9		100.0
623.6		31.0
499.6		48.3
421.7		37.3

200   Cobalt (II) sulfate $CoSO_4 \cdot 7H_2O$

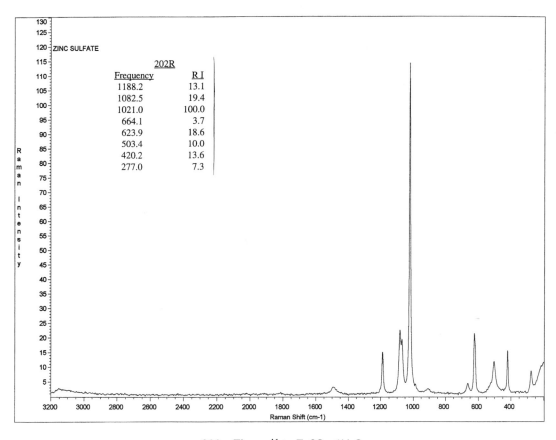

	202R	
Frequency		R I
1188.2		13.1
1082.5		19.4
1021.0		100.0
664.1		3.7
623.9		18.6
503.4		10.0
420.2		13.6
277.0		7.3

202   Zinc sulfate $ZnSO_4 \cdot 6H_2O$

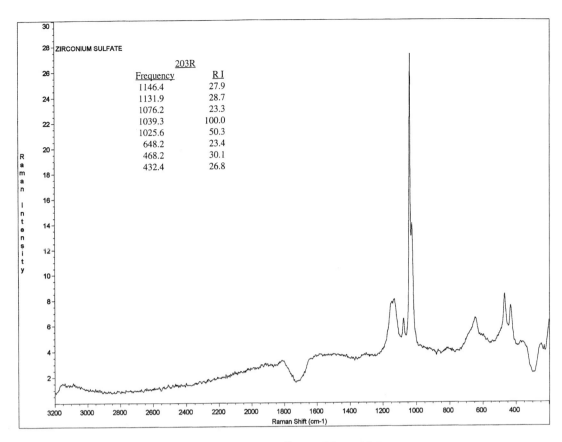

203R	
Frequency	R I
1146.4	27.9
1131.9	28.7
1076.2	23.3
1039.3	100.0
1025.6	50.3
648.2	23.4
468.2	30.1
432.4	26.8

203   Zirconium sulfate $Zr(SO_4)_2 \cdot 4H_2O$

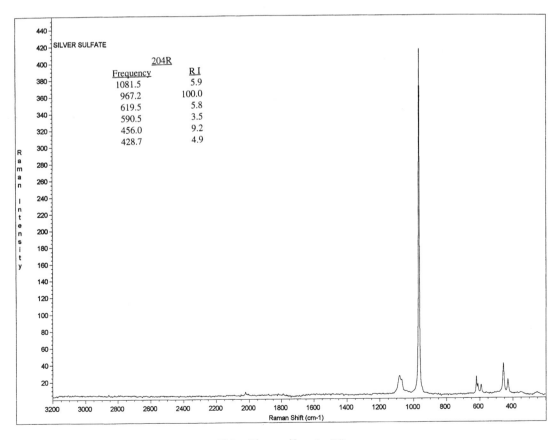

204R	
Frequency	R I
1081.5	5.9
967.2	100.0
619.5	5.8
590.5	3.5
456.0	9.2
428.7	4.9

204   Silver sulfate Ag_2SO_4

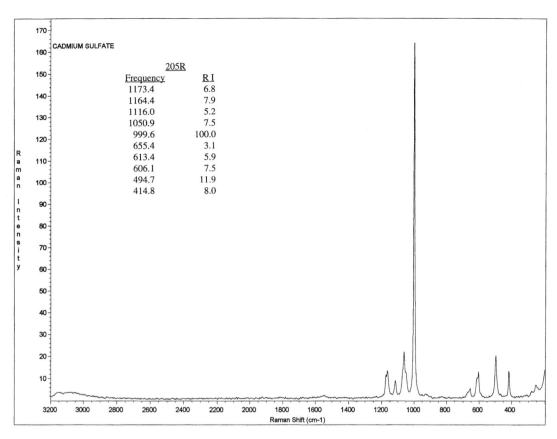

205   Cadmium sulfate CdSO₄·7H₂O

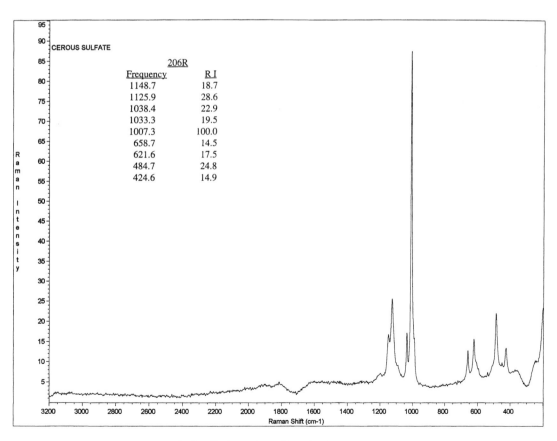

206   Cerium (III) sulfate Ce₂(SO₄)₃·4H₂O

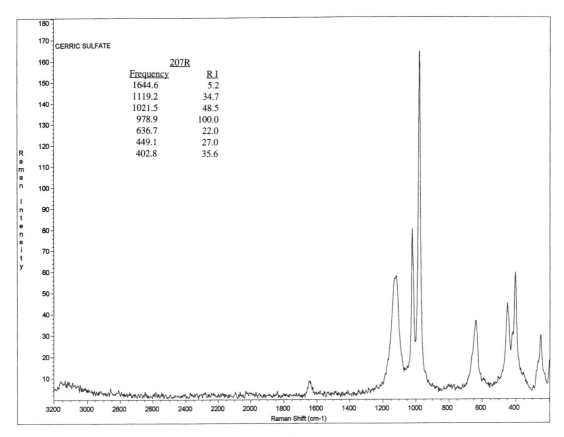

CERRIC SULFATE

207R	
Frequency	R I
1644.6	5.2
1119.2	34.7
1021.5	48.5
978.9	100.0
636.7	22.0
449.1	27.0
402.8	35.6

207   Cerium (IV) sulfate $Ce(SO_4)_2 \cdot 4H_2O$

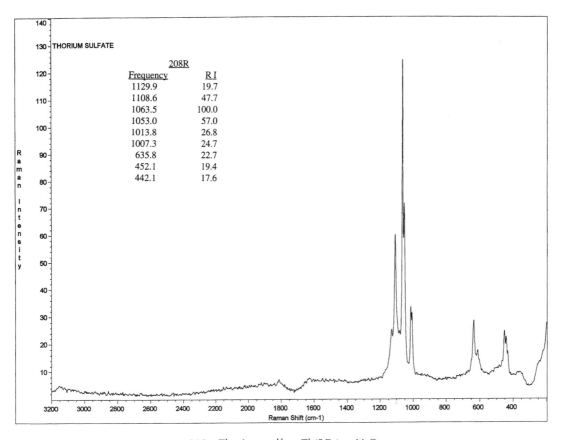

THORIUM SULFATE

208R	
Frequency	R I
1129.9	19.7
1108.6	47.7
1063.5	100.0
1053.0	57.0
1013.8	26.8
1007.3	24.7
635.8	22.7
452.1	19.4
442.1	17.6

208   Thorium sulfate $Th(SO_4)_2 \cdot xH_2O$

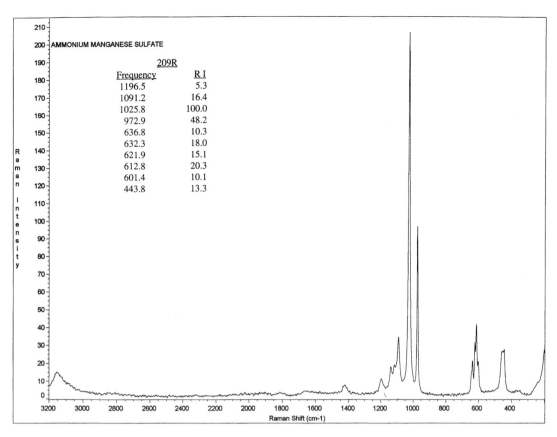

209 Ammonium manganese sulfate $(NH_4)_2MnSO_4 \cdot xH_2O$

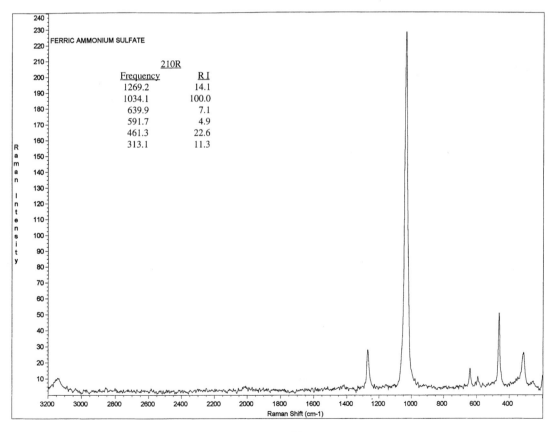

210 Ammonium iron (III) sulfate $(NH_4)Fe(SO_4)_2 \cdot 3H_2O$

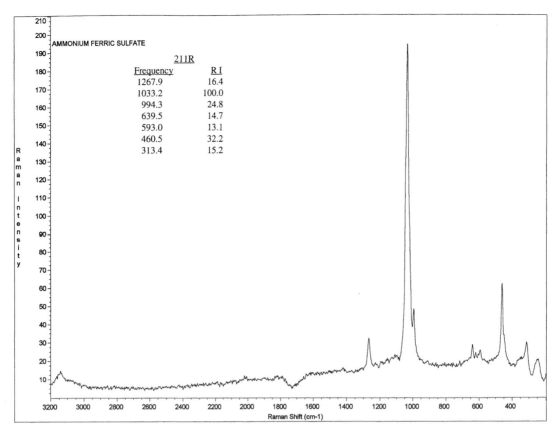

211   Ammonium iron (III) sulfate $NH_4Fe(SO_4)_2 \cdot xH_2O$

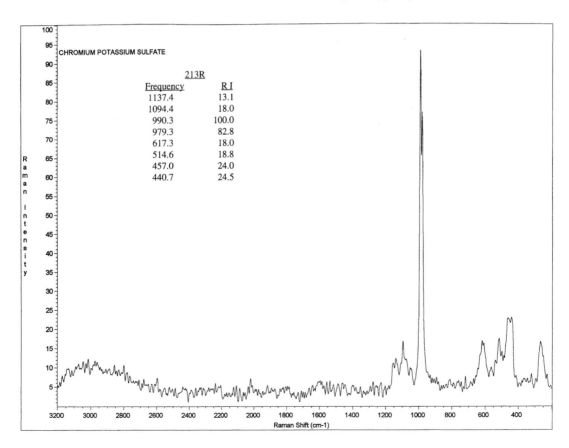

213   Potassium chromium sulfate $KCr(SO_4)_2 \cdot 12H_2O$

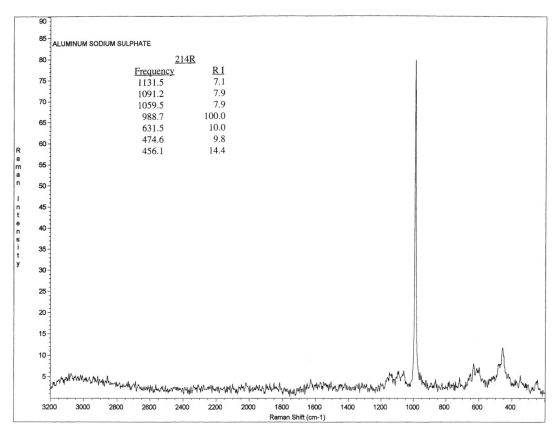

214   Aluminum sodium sulfate $NaAl(SO_4)_2 \cdot xH_2O$

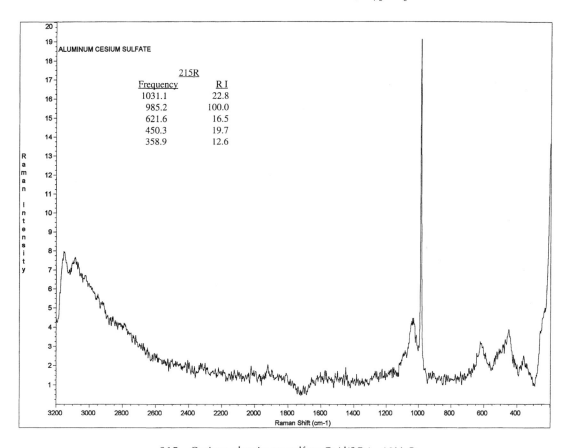

215   Cesium aluminum sulfate $CsAl(SO_4)_2 \cdot 12H_2O$

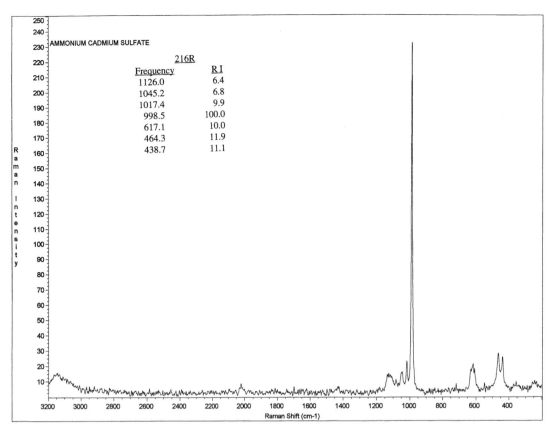

216   Ammonium cadmium sulfate $(NH_4)Cd(SO_4)_2 \cdot 6H_2O$

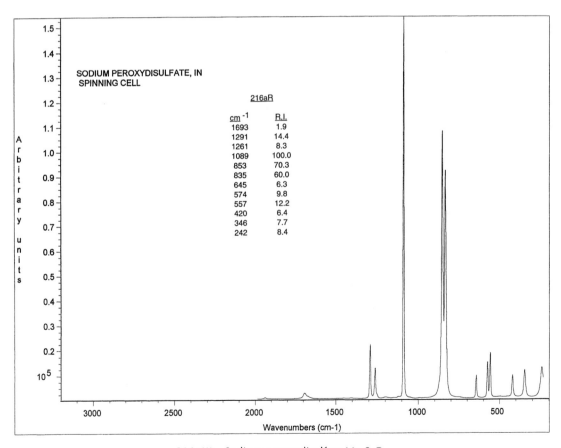

216a(1)   Sodium peroxydisulfate $Na_2S_2O_8$

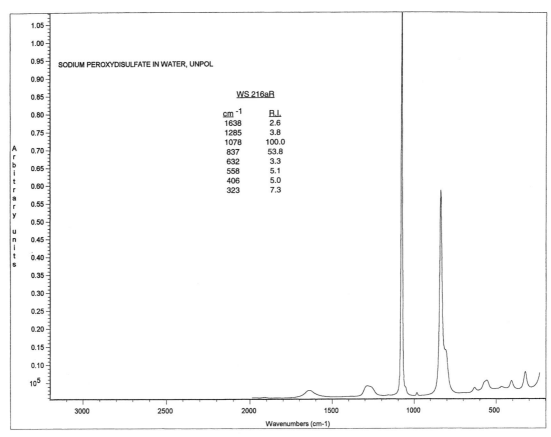

SODIUM PEROXYDISULFATE IN WATER, UNPOL

WS 216aR

cm^{-1}	R.I.
1638	2.6
1285	3.8
1078	100.0
837	53.8
632	3.3
558	5.1
406	5.0
323	7.3

216a(2)   Sodium peroxydisulfate Na$_2$S$_2$O$_8$ in water solution

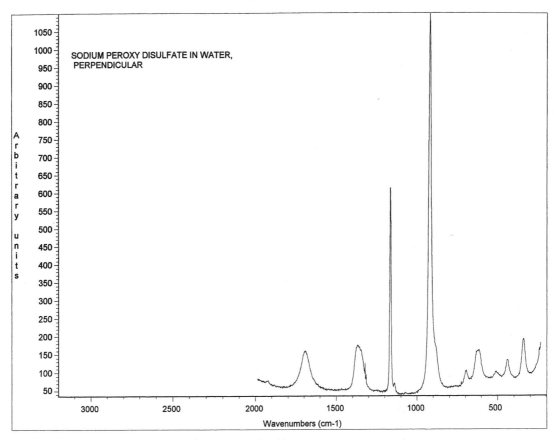

SODIUM PEROXY DISULFATE IN WATER, PERPENDICULAR

216a(3)   Sodium peroxydisulfate Na$_2$S$_2$O$_8$ in water solution

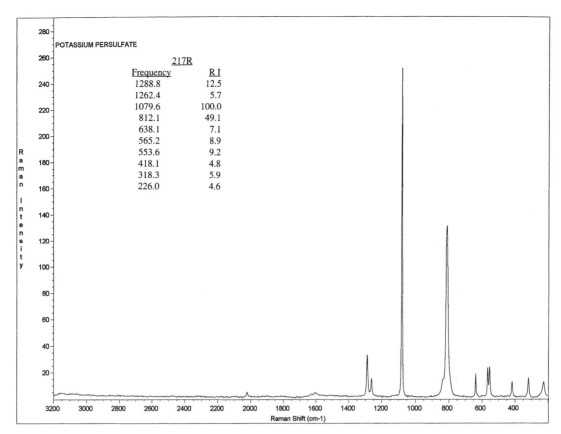

POTASSIUM PERSULFATE

217R

Frequency	R I
1288.8	12.5
1262.4	5.7
1079.6	100.0
812.1	49.1
638.1	7.1
565.2	8.9
553.6	9.2
418.1	4.8
318.3	5.9
226.0	4.6

217   Potassium peroxydisulfate $K_2S_2O_8$

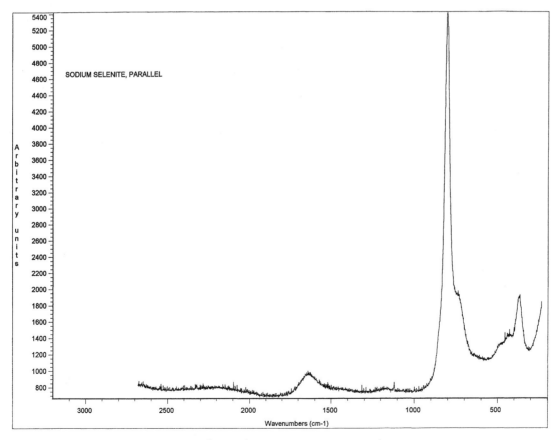

221a   Sodium selenite Na_2SeO_3 in water solution

132

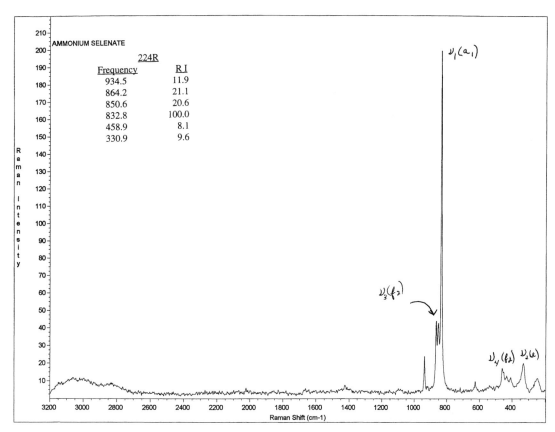

224   Ammonium selenate (NH₄)₂SeO₄

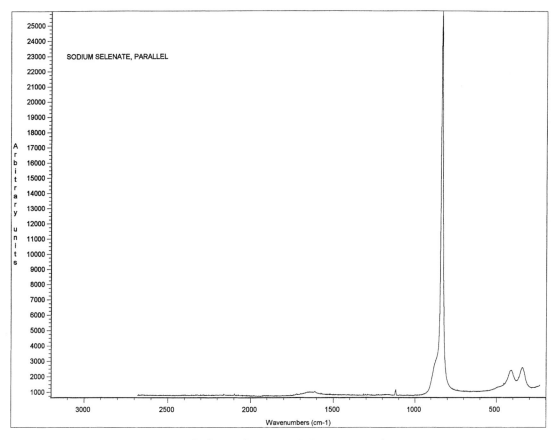

224a   Sodium selenate Na₂SeO₄ in water solution

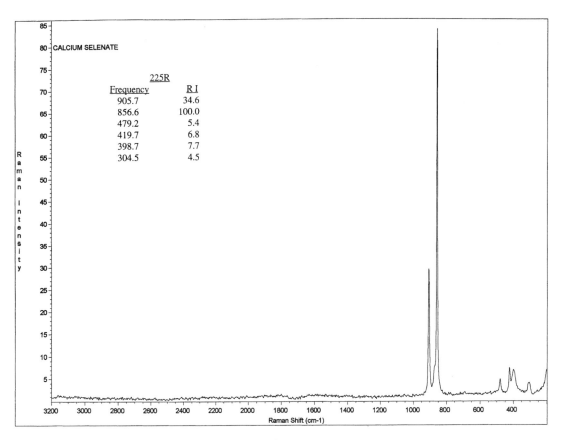

225　Calcium selenate CaSeO$_4$·2H$_2$O

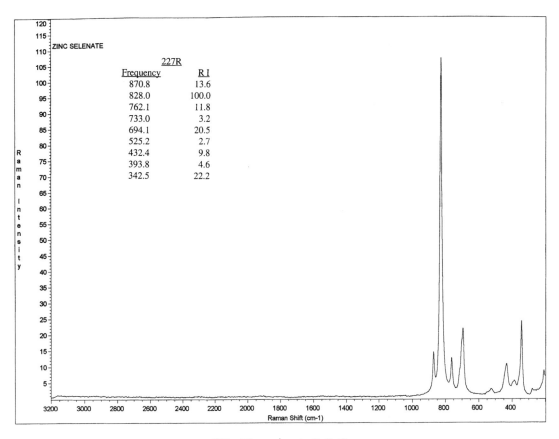

227　Zinc selenate ZnSeO$_4$

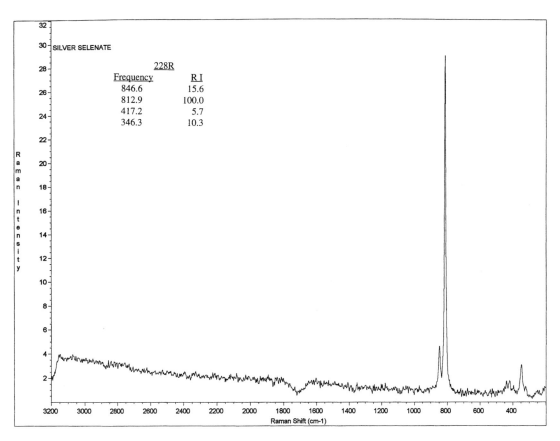

228   Silver selenate $Ag_2SeO_4 \cdot xH_2O$ or wet

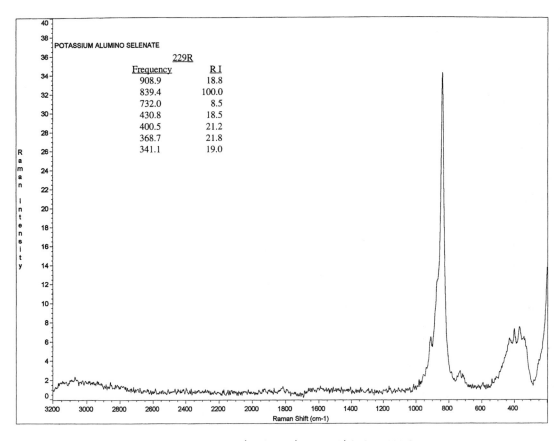

229   Potassium alumino selenate $KAl(SeO_4)_2 \cdot 8H_2O$

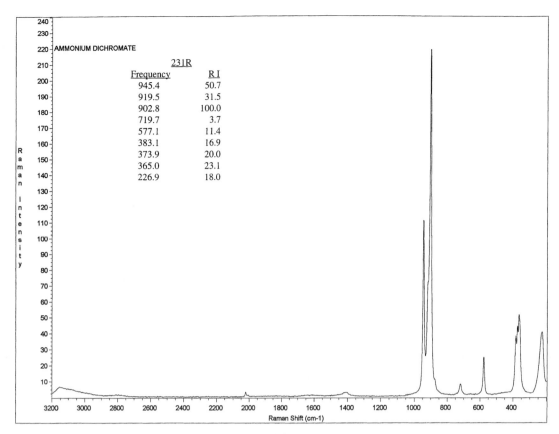

AMMONIUM DICHROMATE

231R

Frequency	R I
945.4	50.7
919.5	31.5
902.8	100.0
719.7	3.7
577.1	11.4
383.1	16.9
373.9	20.0
365.0	23.1
226.9	18.0

231   Ammonium dichromate $(NH_4)_2Cr_2O_7$

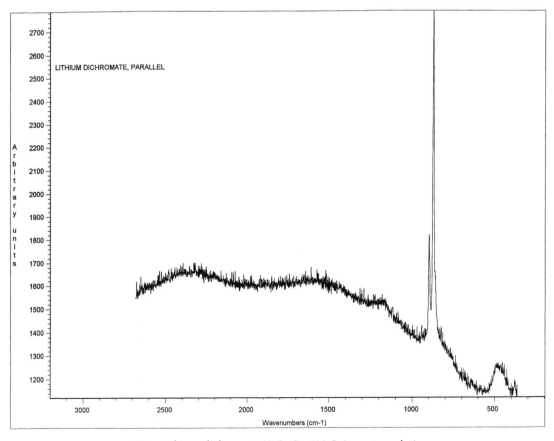

LITHIUM DICHROMATE, PARALLEL

232   Lithium dichromate $Li_2Cr_2O_7 \cdot 2H_2O$ in water solution

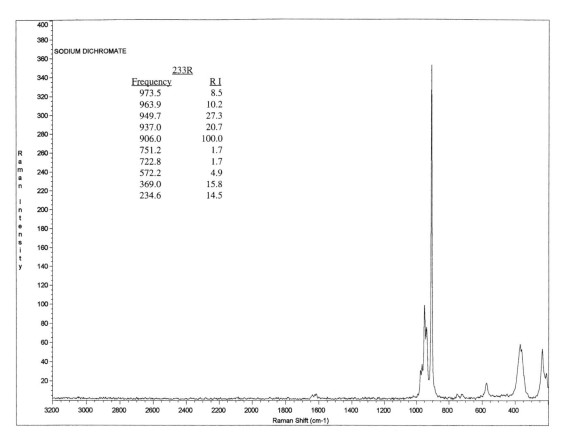

233R	
Frequency	R I
973.5	8.5
963.9	10.2
949.7	27.3
937.0	20.7
906.0	100.0
751.2	1.7
722.8	1.7
572.2	4.9
369.0	15.8
234.6	14.5

233   Sodium dichromate $Na_2Cr_2O_7 \cdot 2H_2O$

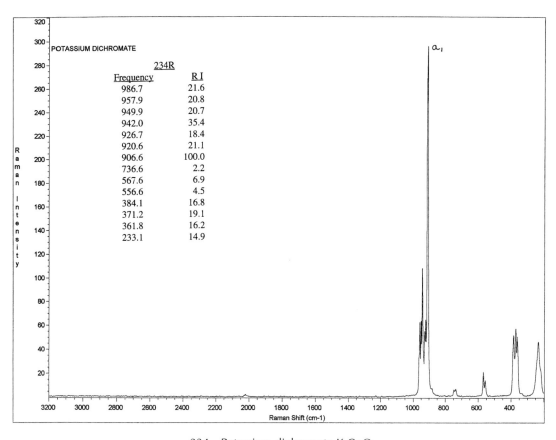

234R	
Frequency	R I
986.7	21.6
957.9	20.8
949.9	20.7
942.0	35.4
926.7	18.4
920.6	21.1
906.6	100.0
736.6	2.2
567.6	6.9
556.6	4.5
384.1	16.8
371.2	19.1
361.8	16.2
233.1	14.9

234   Potassium dichromate $K_2Cr_2O_7$

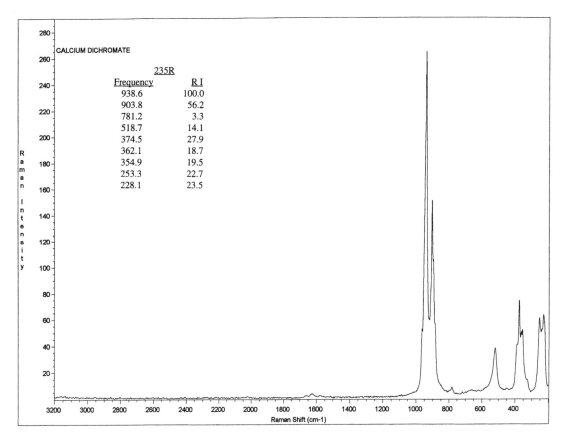

CALCIUM DICHROMATE

235R

Frequency	R I
938.6	100.0
903.8	56.2
781.2	3.3
518.7	14.1
374.5	27.9
362.1	18.7
354.9	19.5
253.3	22.7
228.1	23.5

235   Calcium dichromate $CaCr_2O_7 \cdot xH_2O$

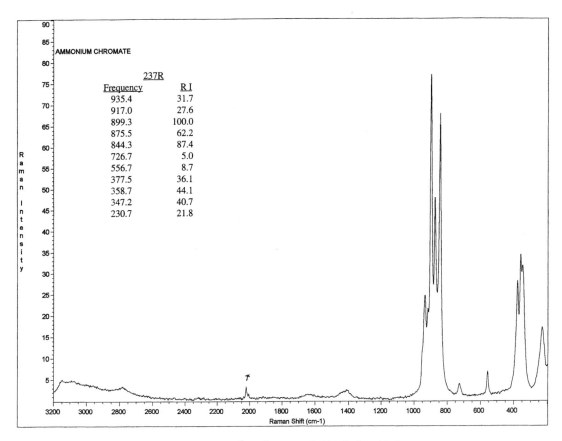

AMMONIUM CHROMATE

237R

Frequency	R I
935.4	31.7
917.0	27.6
899.3	100.0
875.5	62.2
844.3	87.4
726.7	5.0
556.7	8.7
377.5	36.1
358.7	44.1
347.2	40.7
230.7	21.8

237   Ammonium chromate $(NH_4)_2CrO_4 \cdot xH_2O$

138

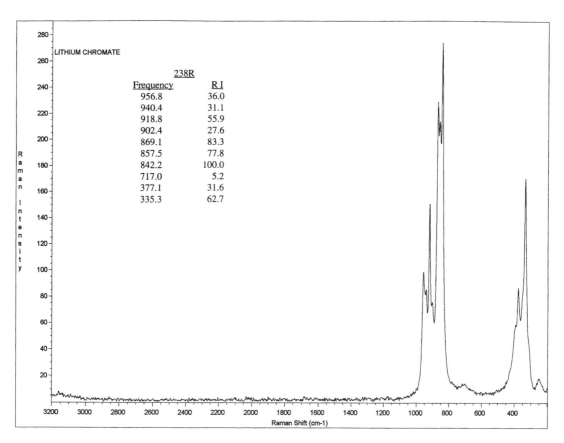

238R	
Frequency	R I
956.8	36.0
940.4	31.1
918.8	55.9
902.4	27.6
869.1	83.3
857.5	77.8
842.2	100.0
717.0	5.2
377.1	31.6
335.3	62.7

LITHIUM CHROMATE

238   Lithium chromate $Li_2CrO_4 \cdot xH_2O$

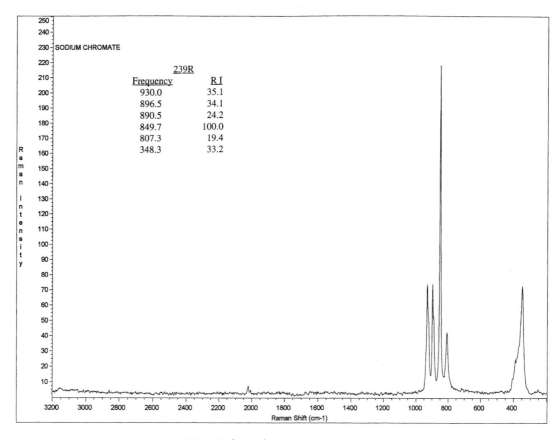

239R	
Frequency	R I
930.0	35.1
896.5	34.1
890.5	24.2
849.7	100.0
807.3	19.4
348.3	33.2

SODIUM CHROMATE

239   Sodium chromate $Na_2CrO_4 \cdot xH_2O$

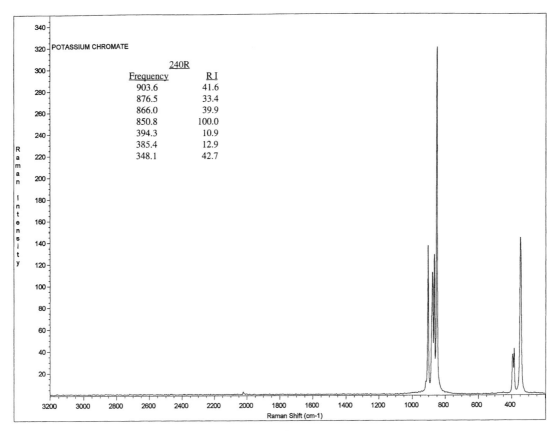

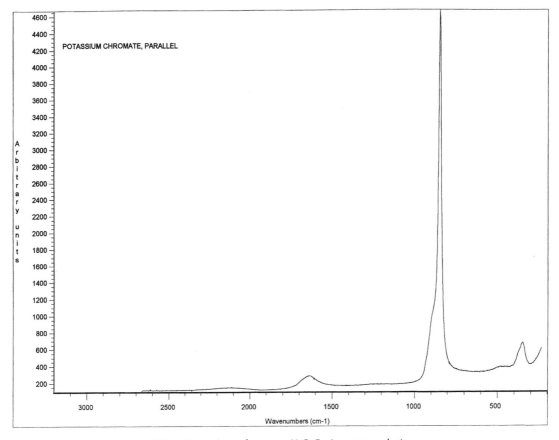

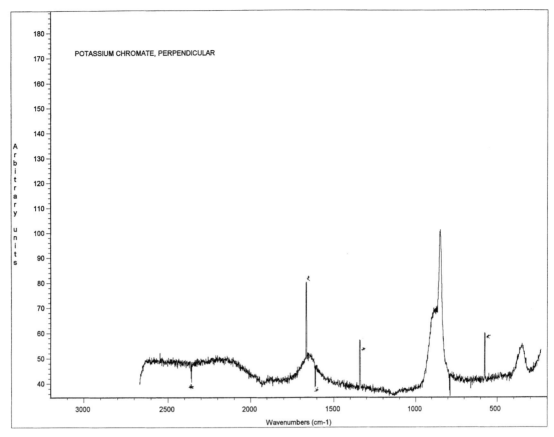

240b   Potassium chromate K₂CrO₄ in water solution

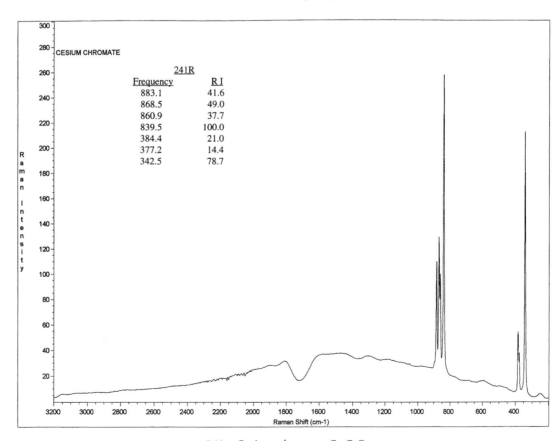

241   Cesium chromate Cs₂CrO₄

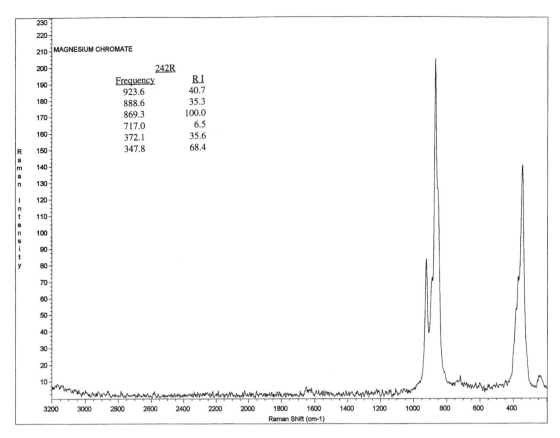

242   Magnesium chromate MgCrO$_4$

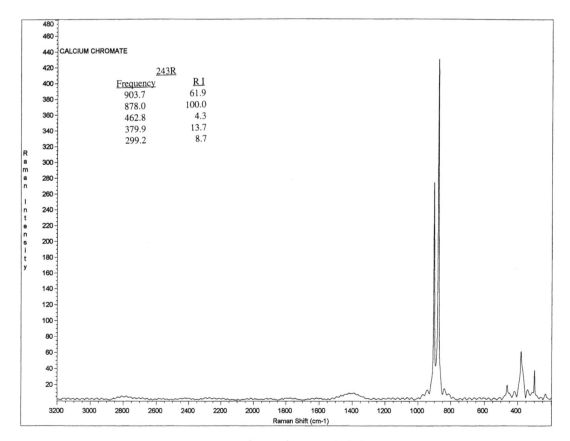

243   Calcium chromate CaCrO$_4$

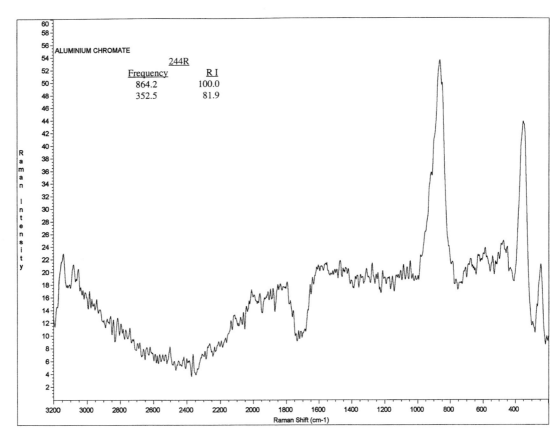

244　Aluminum chromate Al$_2$(CrO$_4$)$_3 \cdot$xH$_2$O

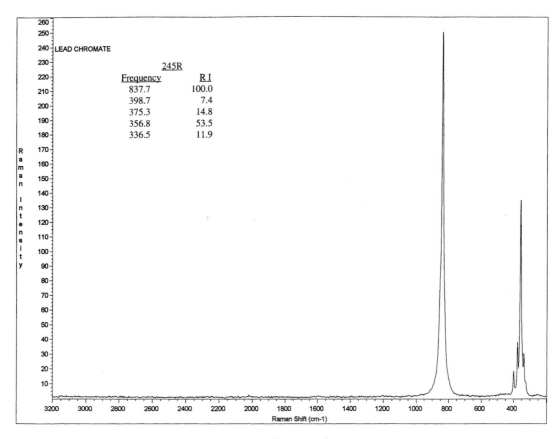

245　Lead chromate PbCrO$_4$

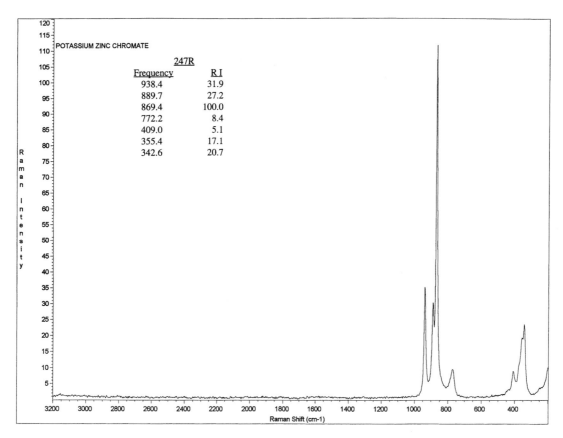

POTASSIUM ZINC CHROMATE

247R	
Frequency	R I
938.4	31.9
889.7	27.2
869.4	100.0
772.2	8.4
409.0	5.1
355.4	17.1
342.6	20.7

247   Potassium zinc chromate $K_2CrO_4 \cdot 3ZnCrO_4 \cdot Zn(OH)_2$

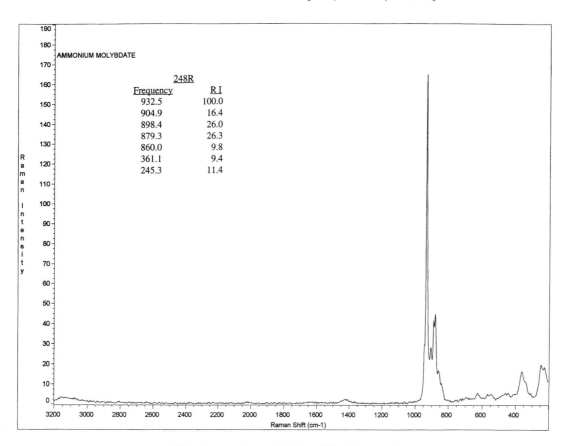

AMMONIUM MOLYBDATE

248R	
Frequency	R I
932.5	100.0
904.9	16.4
898.4	26.0
879.3	26.3
860.0	9.8
361.1	9.4
245.3	11.4

248   Ammonium molybdate (VI) $(NH_4)_2MoO_4$

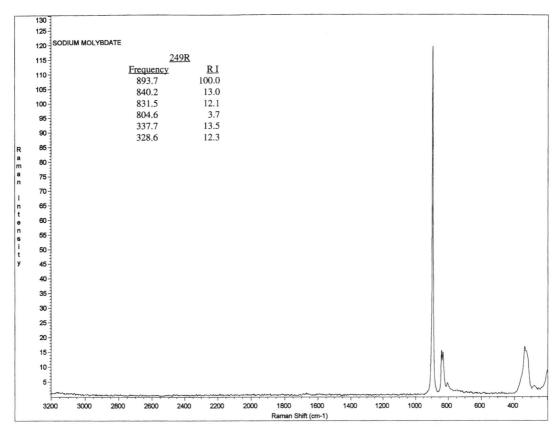

| 249R | |
Frequency	R I
893.7	100.0
840.2	13.0
831.5	12.1
804.6	3.7
337.7	13.5
328.6	12.3

SODIUM MOLYBDATE

249   Sodium molybdate (VI) Na$_2$MoO$_4 \cdot$2H$_2$O

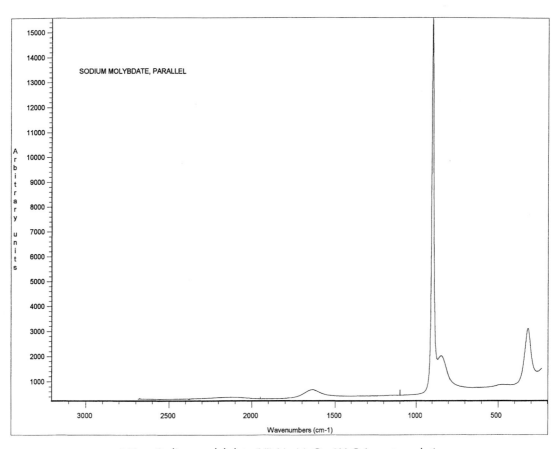

SODIUM MOLYBDATE, PARALLEL

249a   Sodium molybdate (VI) Na$_2$MoO$_4 \cdot$2H$_2$O in water solution

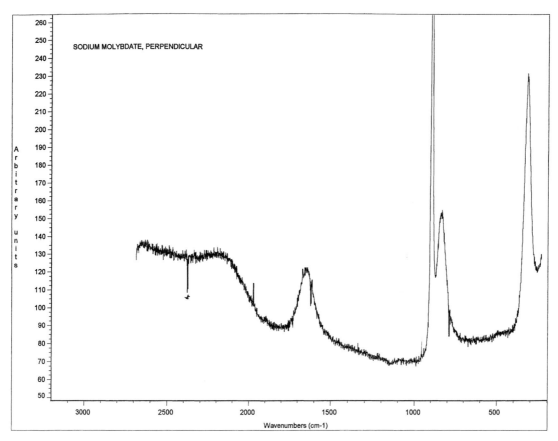

249b   Sodium molybdate (VI) $Na_2MoO_4 \cdot 2H_2O$ in water solution

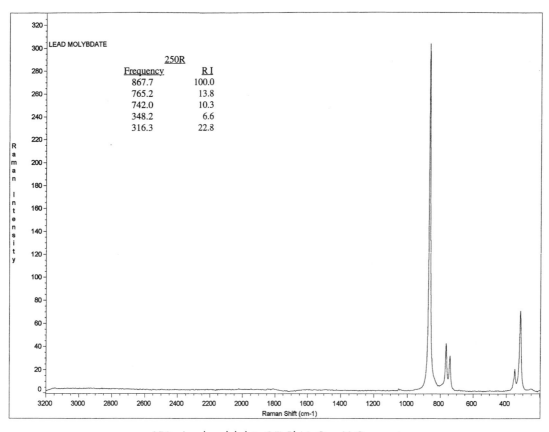

250   Lead molybdate (VI) $PbMoO_4 \cdot xH_2O$ or wet

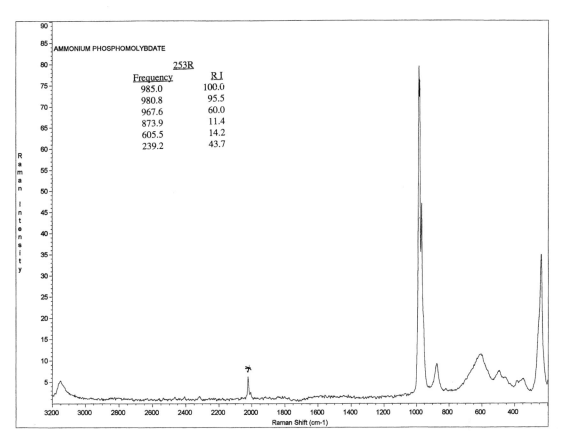

	253R	
Frequency		RI
985.0		100.0
980.8		95.5
967.6		60.0
873.9		11.4
605.5		14.2
239.2		43.7

253   Ammonium phosphomolybdate $(NH_4)_3PMo_{12}O_{40}$

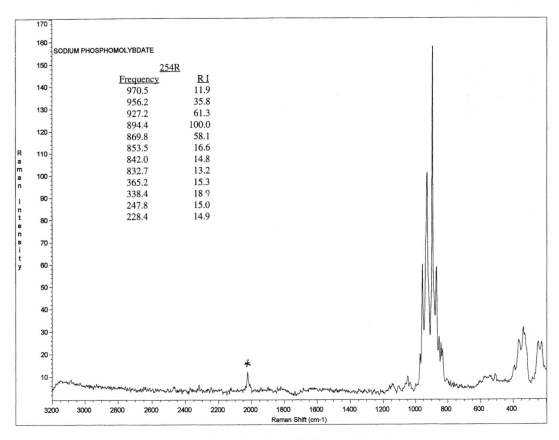

	254R	
Frequency		RI
970.5		11.9
956.2		35.8
927.2		61.3
894.4		100.0
869.8		58.1
853.5		16.6
842.0		14.8
832.7		13.2
365.2		15.3
338.4		18.9
247.8		15.0
228.4		14.9

254   Sodium phosphomolybdate $Na_3PMo_{12}O_{40}$

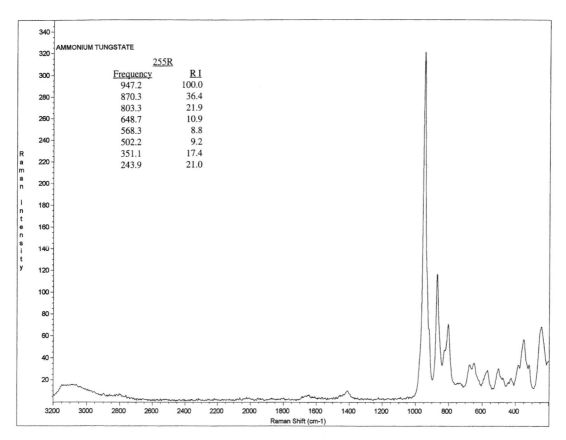

255　Ammonium tungstate $(NH_4)_2WO_4$

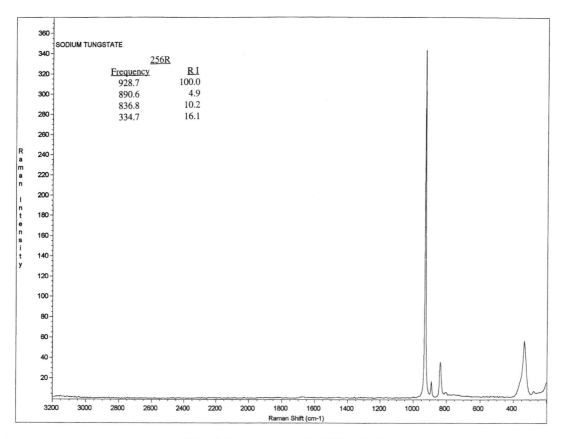

256　Sodium tungstate $Na_2WO_4 \cdot 2H_2O$

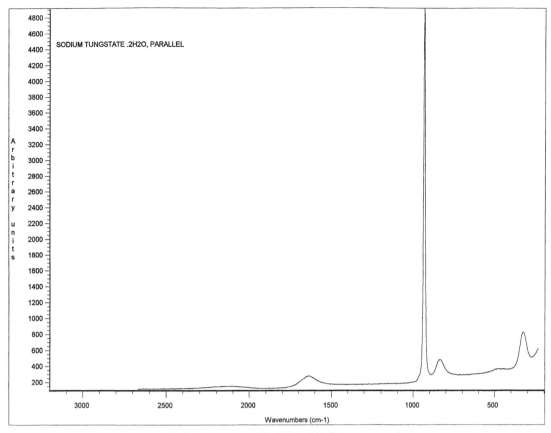

256a Sodium tungstate $Na_2WO_4 \cdot 2H_2O$ in water solution

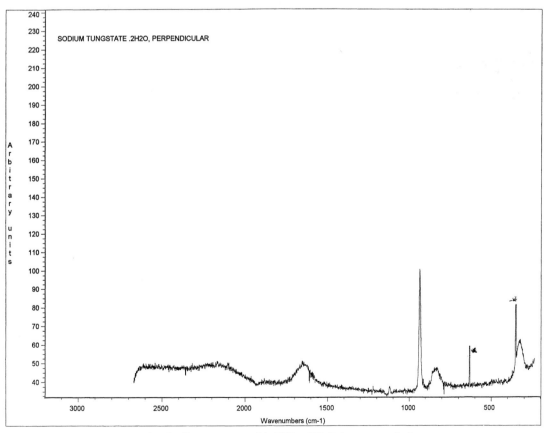

256b Sodium tungstate $Na_2WO_4 \cdot 2H_2O$ in water solution

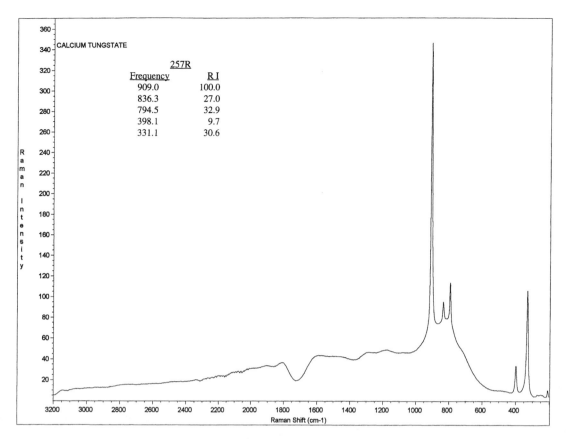

257   Calcium tungstate CaWO$_4$

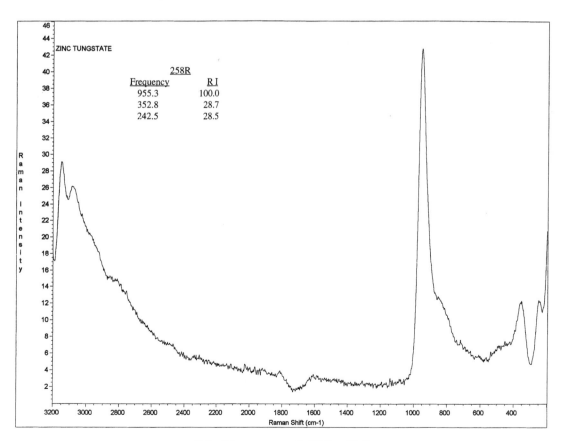

258   Zinc tungstate ZnWO$_4$·xH$_2$O

150

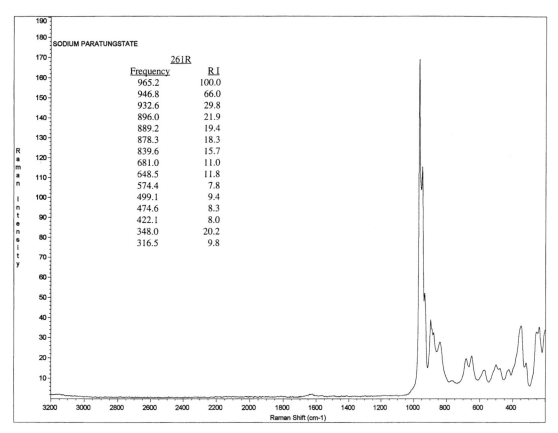

261R	
Frequency	R I
965.2	100.0
946.8	66.0
932.6	29.8
896.0	21.9
889.2	19.4
878.3	18.3
839.6	15.7
681.0	11.0
648.5	11.8
574.4	7.8
499.1	9.4
474.6	8.3
422.1	8.0
348.0	20.2
316.5	9.8

261   Sodium paratungstate $Na_6W_7O_{24} \cdot 16H_2O$

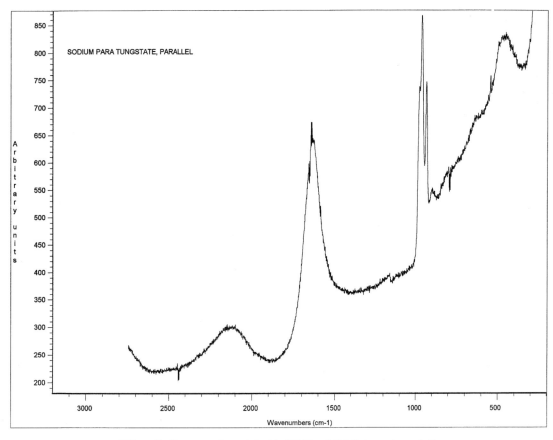

261a   Sodium paratungstate $Na_6W_7O_{24} \cdot 16H_2O$ in water solution

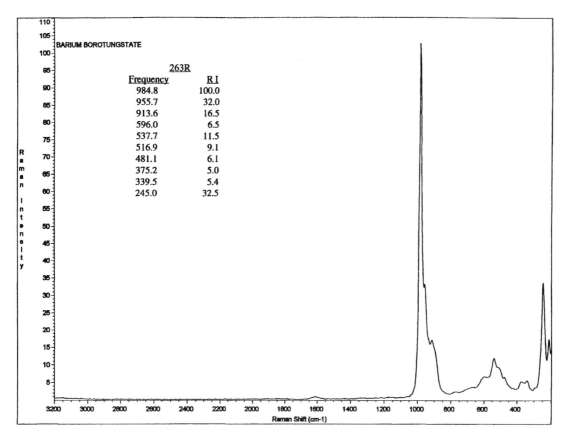

263R	
Frequency	**R I**
984.8	100.0
955.7	32.0
913.6	16.5
596.0	6.5
537.7	11.5
516.9	9.1
481.1	6.1
375.2	5.0
339.5	5.4
245.0	32.5

263   Barium borotungstate $Ba_3(BW_{12}O_{40})_2 \cdot xH_2O$

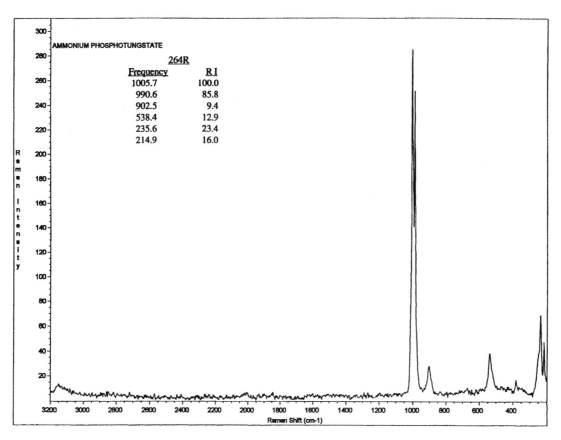

AMMONIUM PHOSPHOTUNGSTATE

264R	
Frequency	**R I**
1005.7	100.0
990.6	85.8
902.5	9.4
538.4	12.9
235.6	23.4
214.9	16.0

264   Ammonium phosphotungstate $(NH_4)_3PW_{12}O_{40} \cdot 4H_2O$

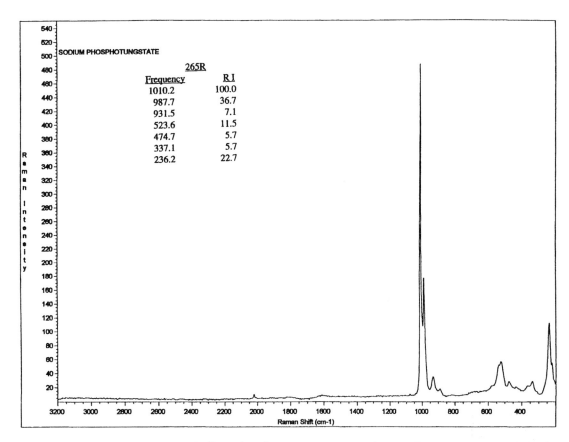

265   Sodium phosphotungstate $Na_3PW_{12}O_{40} \cdot xH_2O$

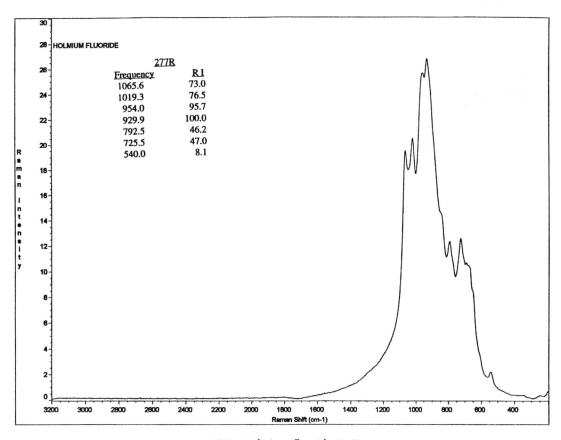

277   Holmium fluoride HoF_3

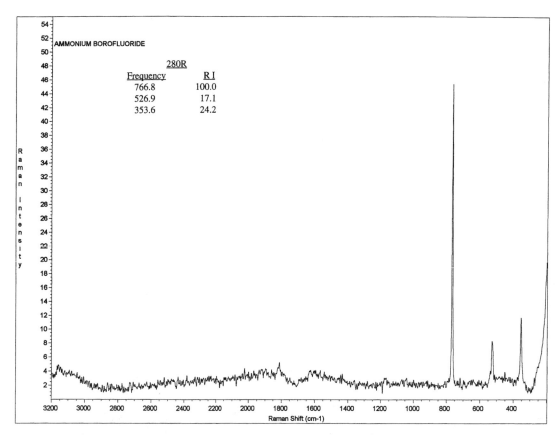

AMMONIUM BOROFLUORIDE

280R

Frequency	R I
766.8	100.0
526.9	17.1
353.6	24.2

280   Ammonium tetrafluoroborate NH_4BF_4

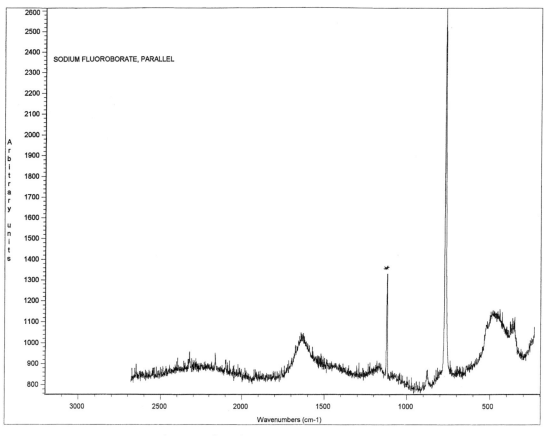

SODIUM FLUOROBORATE, PARALLEL

281   Sodium tetrafluoroborate $NaBF_4 \cdot xH_2O$ in water solution

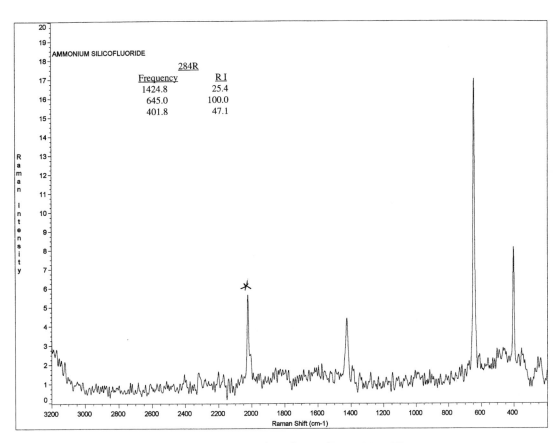

284   Ammonium hexafluorosilicate $(NH_4)_2SiF_6$

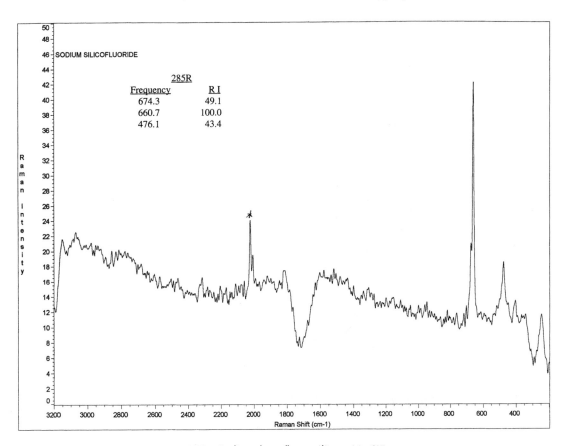

285   Sodium hexafluorosilicate Na_2SiF_6

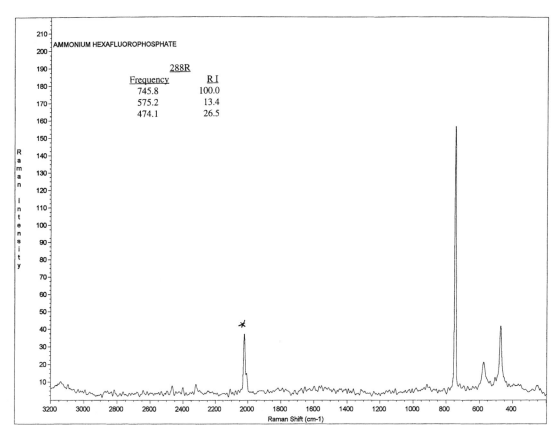

288   Ammonium hexafluorophosphate NH$_4$PF$_6$

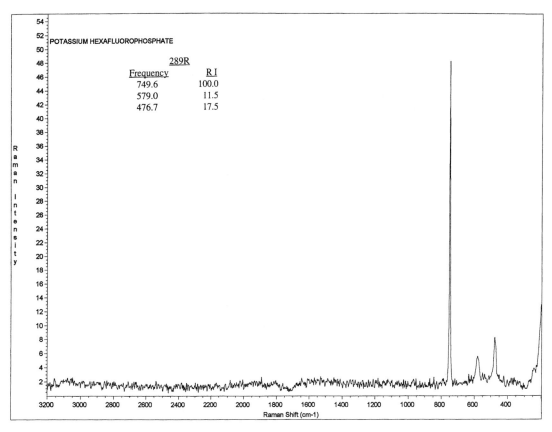

289   Potassium hexafluorophosphate KPF$_6$·xH$_2$O or wet

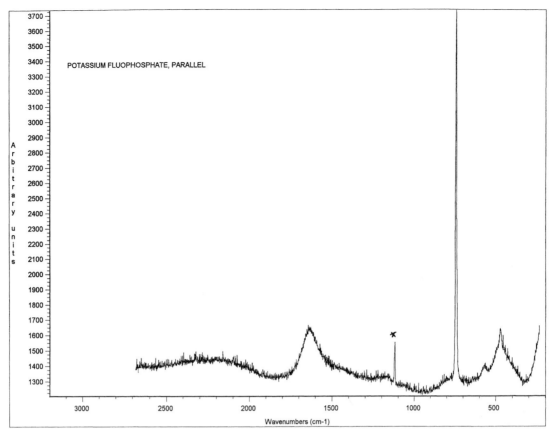

289   Potassium hexafluorophosphate $KPF_6 \cdot xH_2O$ or wet in water solution

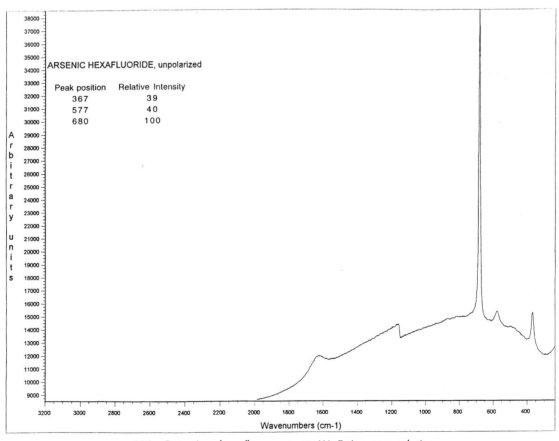

290   Potassium hexafluoroarsenate $KAsF_6$ in water solution

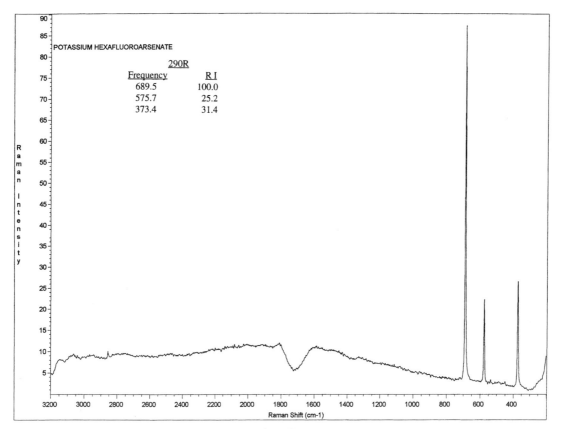

POTASSIUM HEXAFLUOROARSENATE

290R

Frequency	R I
689.5	100.0
575.7	25.2
373.4	31.4

290a   Potassium hexafluoroarsenate KAsF$_6$

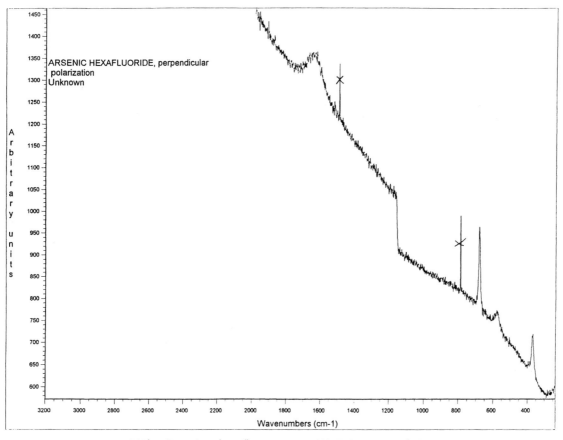

ARSENIC HEXAFLUORIDE, perpendicular
polarization
Unknown

290b   Potassium hexafluoroarsenate KAsF$_6$ in water solution

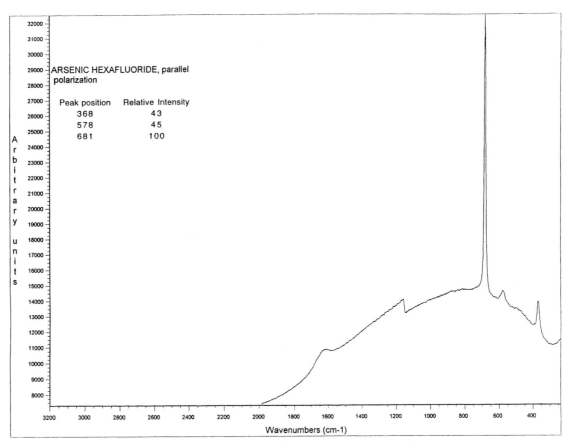

ARSENIC HEXAFLUORIDE, parallel polarization

Peak position	Relative Intensity
368	43
578	45
681	100

290c   Potassium hexafluoroarsenate KAsF$_6$ in water solution

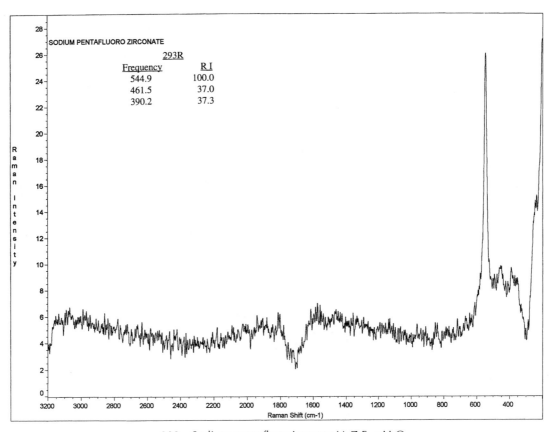

SODIUM PENTAFLUORO ZIRCONATE

293R

Frequency	RI
544.9	100.0
461.5	37.0
390.2	37.3

293   Sodium pentaflurozirconate NaZrF$_5 \cdot$xH$_2$O

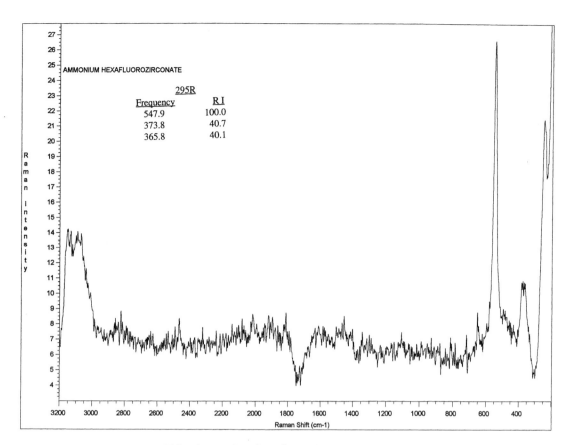

295   Ammonium hexafluorozirconate $(NH_4)_2ZrF_6$

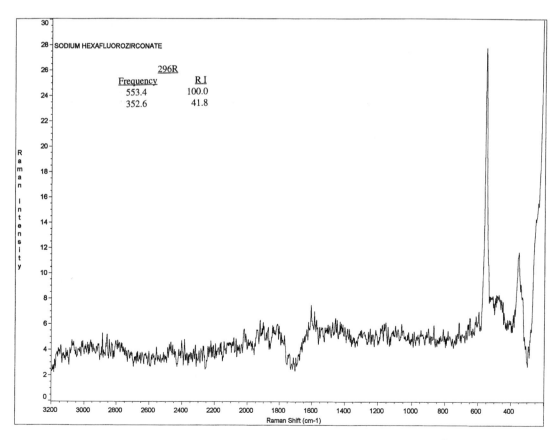

296   Sodium hexafluorozirconate Na_2ZrF_6

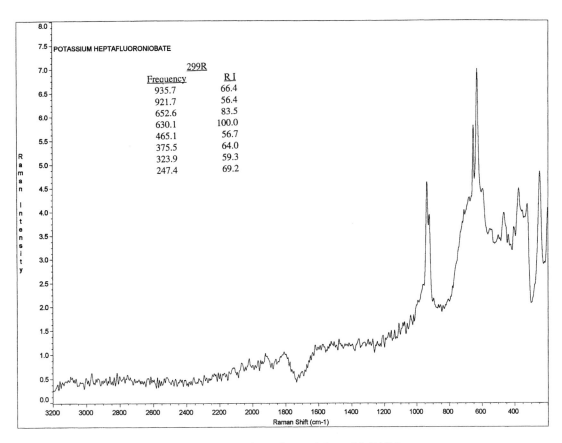

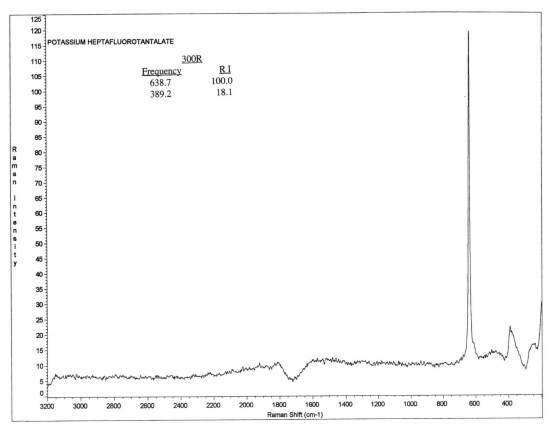

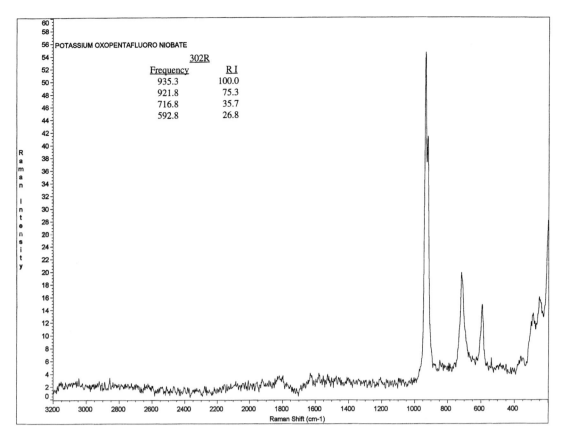

302 Potassium oxopentafluoroniobate $K_2NbOF_5 \cdot xH_2O$

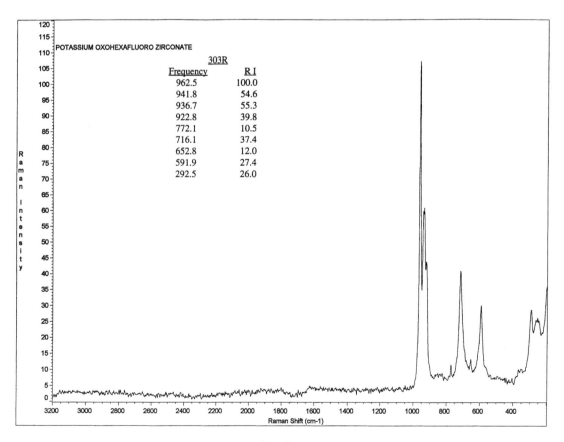

303 Potassium oxohexafluorozirconate K_3ZrOF_6

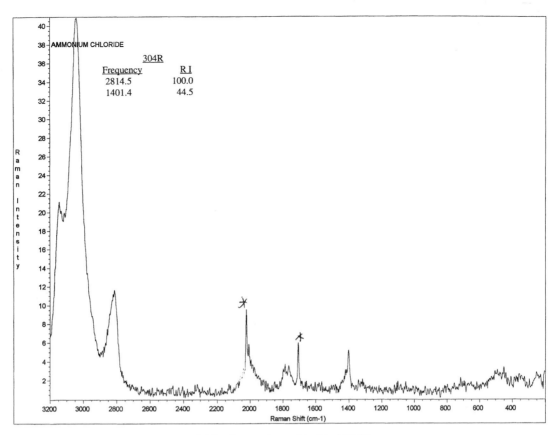

304   Ammonium chloride NH₄Cl

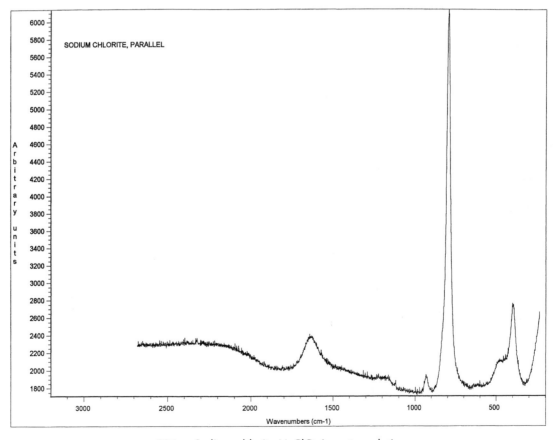

304a   Sodium chlorite NaClO₂ in water solution

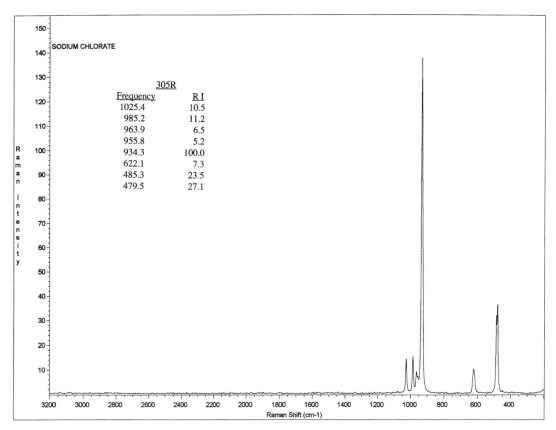

305R	
Frequency	R I
1025.4	10.5
985.2	11.2
963.9	6.5
955.8	5.2
934.3	100.0
622.1	7.3
485.3	23.5
479.5	27.1

SODIUM CHLORATE

305   Sodium chlorate NaClO₃

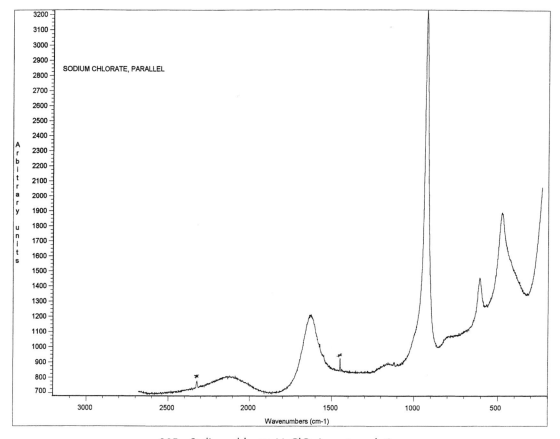

SODIUM CHLORATE, PARALLEL

305   Sodium chlorate NaClO₃ in water solution

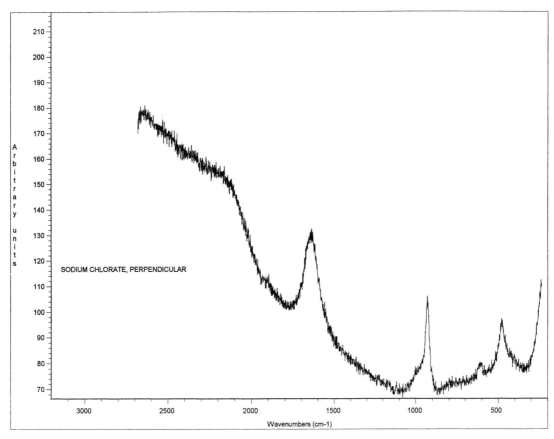

SODIUM CHLORATE, PERPENDICULAR

305a   Sodium chlorate NaClO$_3$ in water solution

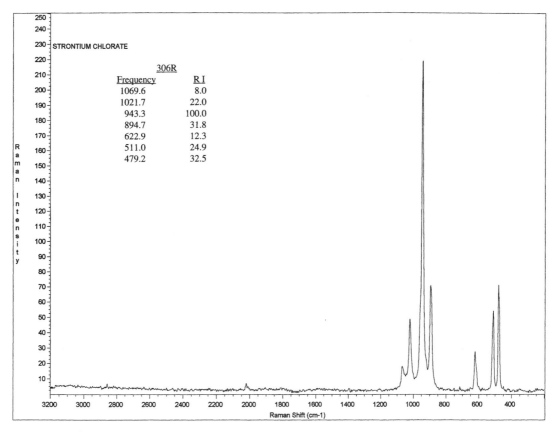

STRONTIUM CHLORATE

306R	
Frequency	R I
1069.6	8.0
1021.7	22.0
943.3	100.0
894.7	31.8
622.9	12.3
511.0	24.9
479.2	32.5

306   Strontium chlorate Sr(ClO$_3$)$_2$·xH$_2$O

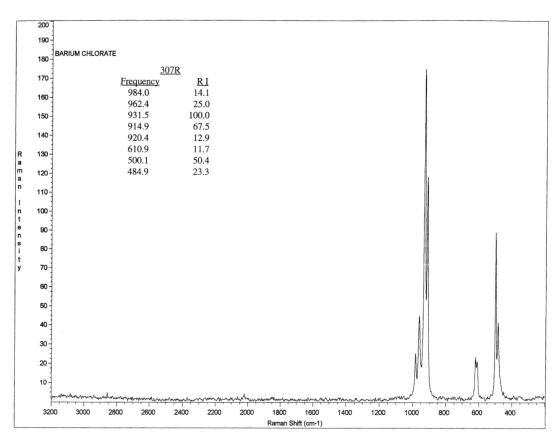

307   Barium chlorate Ba(ClO$_3$)$_2$·H$_2$O

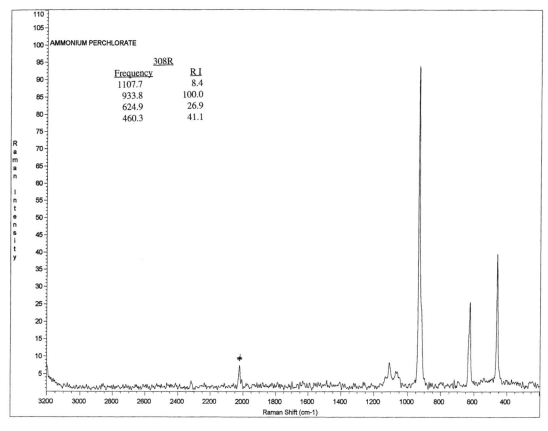

308   Ammonium perchlorate NH$_4$ClO$_4$

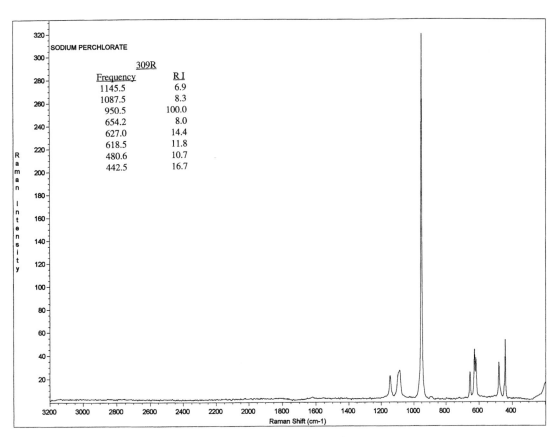

SODIUM PERCHLORATE

309R

Frequency	R I
1145.5	6.9
1087.5	8.3
950.5	100.0
654.2	8.0
627.0	14.4
618.5	11.8
480.6	10.7
442.5	16.7

309   Sodium perchlorate $NaClO_4 \cdot H_2O$

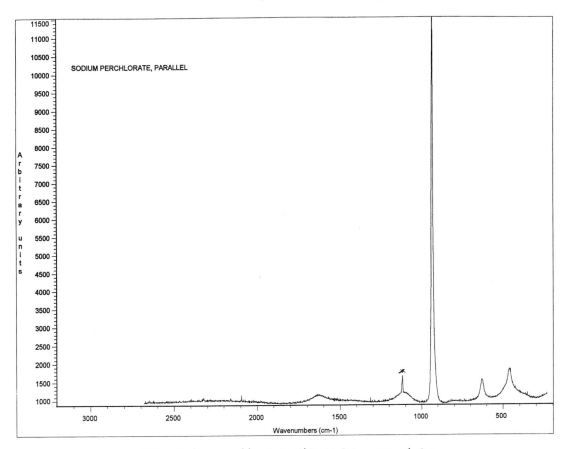

SODIUM PERCHLORATE, PARALLEL

309a   Sodium perchlorate $NaClO_4 \cdot H_2O$ in water solution

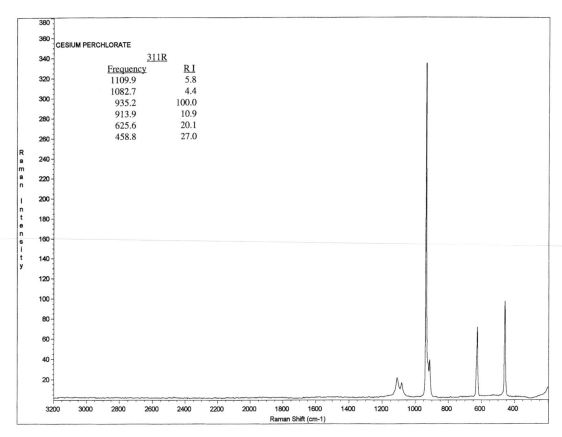

311   Cesium perchlorate CsClO$_4$

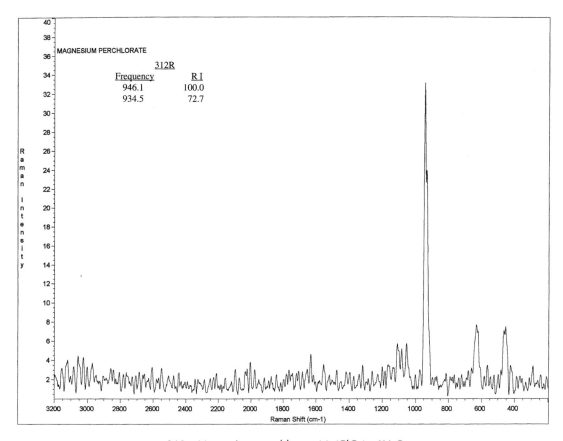

312   Magnesium perchlorate Mg(ClO$_4$)$_2 \cdot$ 6H$_2$O

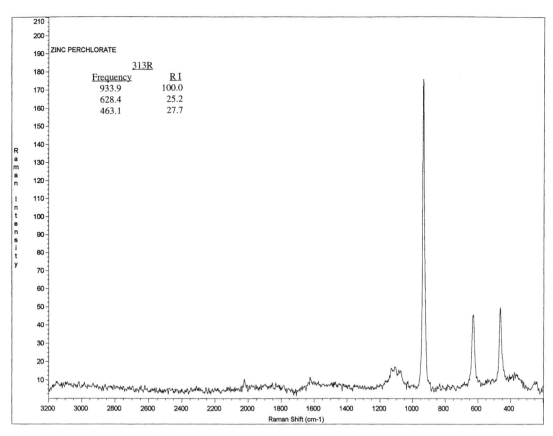

313   Zinc perchlorate Zn(ClO$_4$)$_2$·6H$_2$O

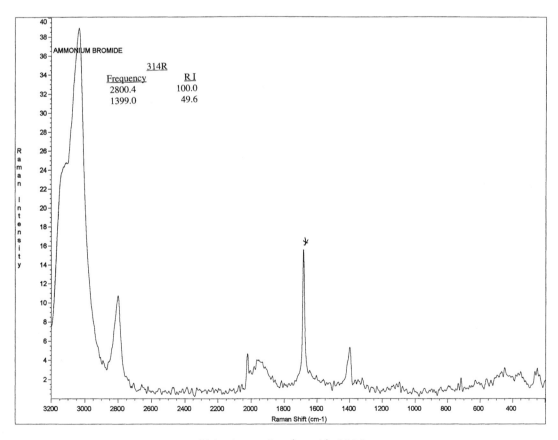

314   Ammonium bromide NH$_4$Br

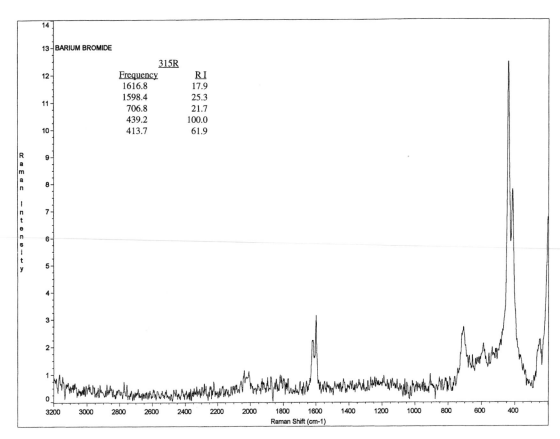

315   Barium bromide BaBr₂·2H₂O

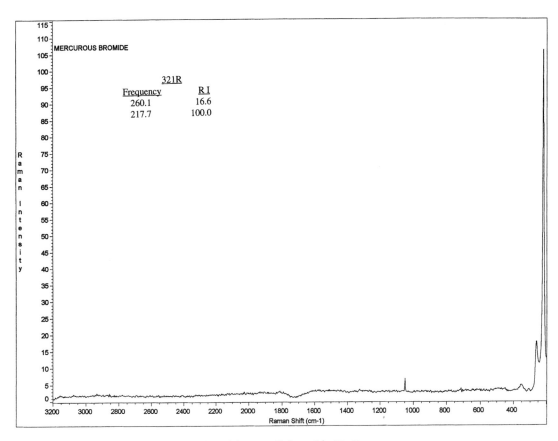

321   Mercury (I) bromide Hg₂Br₂

170

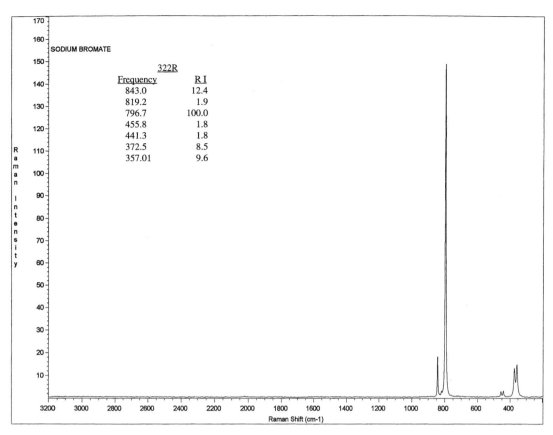

322R	
Frequency	R I
843.0	12.4
819.2	1.9
796.7	100.0
455.8	1.8
441.3	1.8
372.5	8.5
357.01	9.6

SODIUM BROMATE

322   Sodium bromate NaBrO$_3$

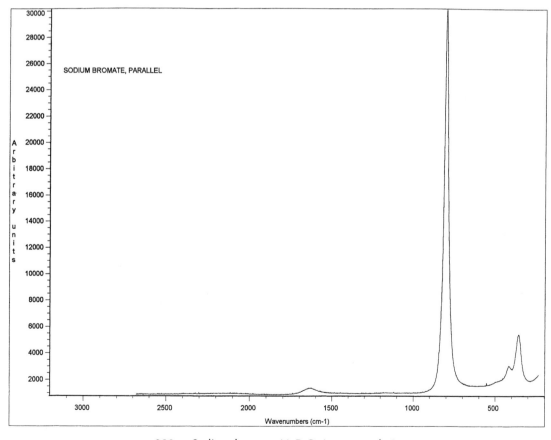

SODIUM BROMATE, PARALLEL

322a   Sodium bromate NaBrO$_3$ in water solution

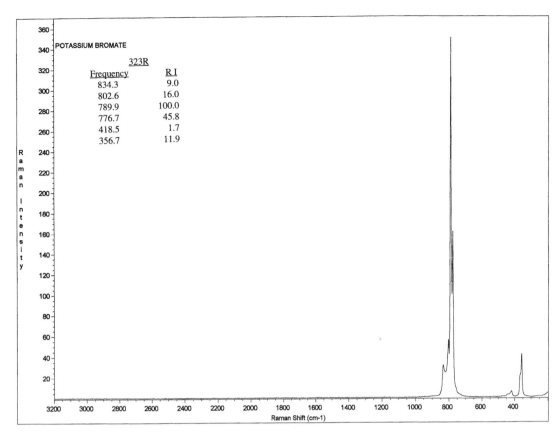

POTASSIUM BROMATE

323R

Frequency	R I
834.3	9.0
802.6	16.0
789.9	100.0
776.7	45.8
418.5	1.7
356.7	11.9

323   Potassium bromate KBrO$_3$

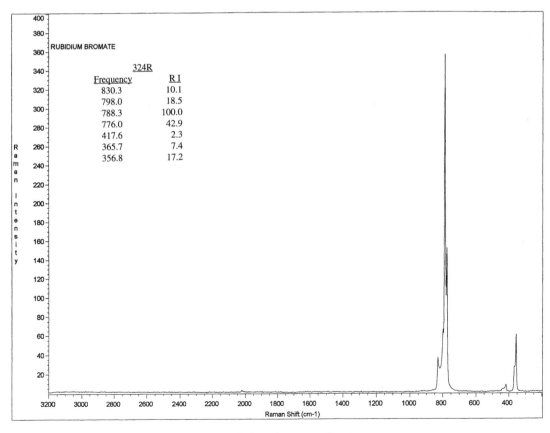

RUBIDIUM BROMATE

324R

Frequency	R I
830.3	10.1
798.0	18.5
788.3	100.0
776.0	42.9
417.6	2.3
365.7	7.4
356.8	17.2

324   Rubidium bromate RbBrO$_3$

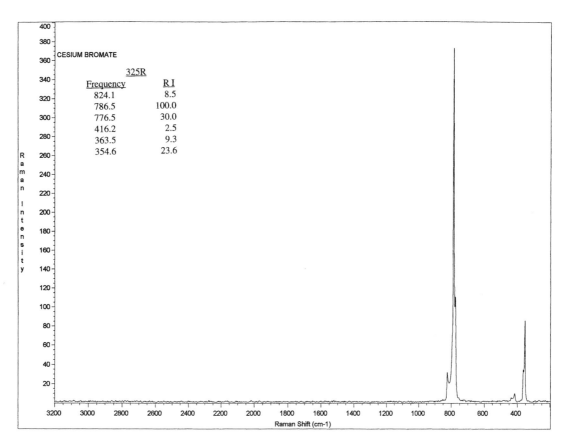

CESIUM BROMATE

325R

Frequency	R I
824.1	8.5
786.5	100.0
776.5	30.0
416.2	2.5
363.5	9.3
354.6	23.6

325   Cesium bromate CsBrO$_3$

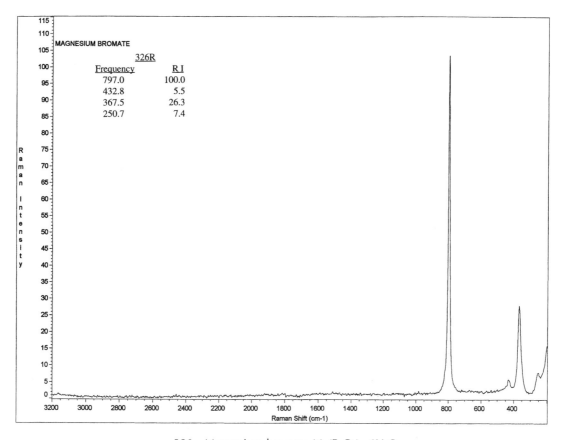

MAGNESIUM BROMATE

326R

Frequency	R I
797.0	100.0
432.8	5.5
367.5	26.3
250.7	7.4

326   Magnesium bromate Mg(BrO$_3$)$_2$·6H$_2$O

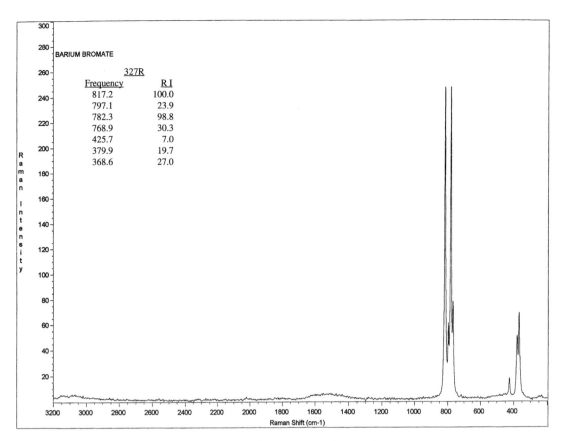

327   Barium bromate Ba(BrO$_3$)$_2 \cdot$H$_2$O

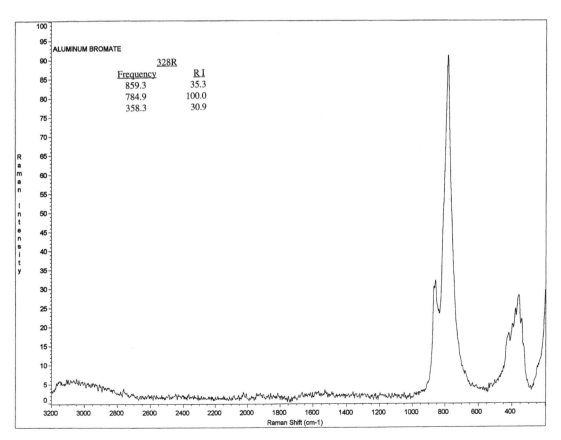

328   Aluminum bromate Al(BrO$_3$)$_3 \cdot$9H$_2$O

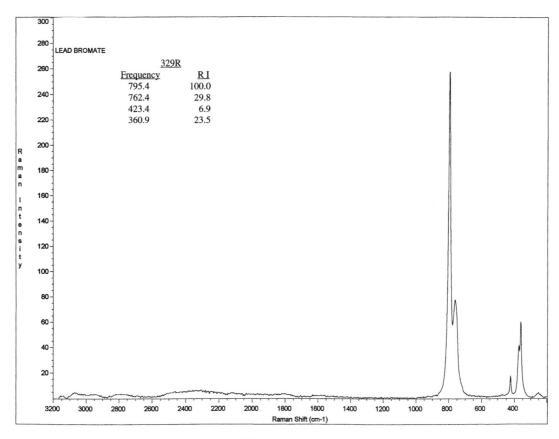

329   Lead bromate $Pb(BrO_3)_2 \cdot H_2O$

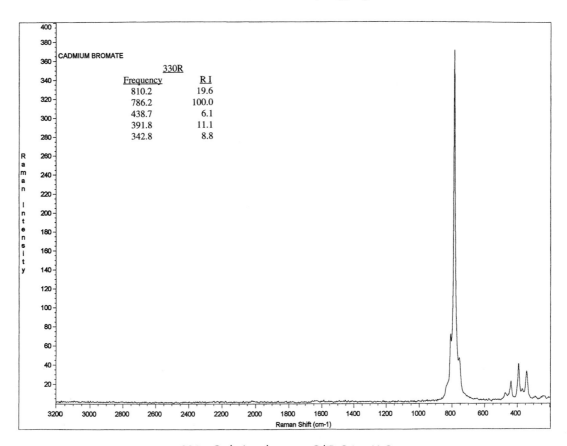

330   Cadmium bromate $Cd(BrO_3)_2 \cdot xH_2O$

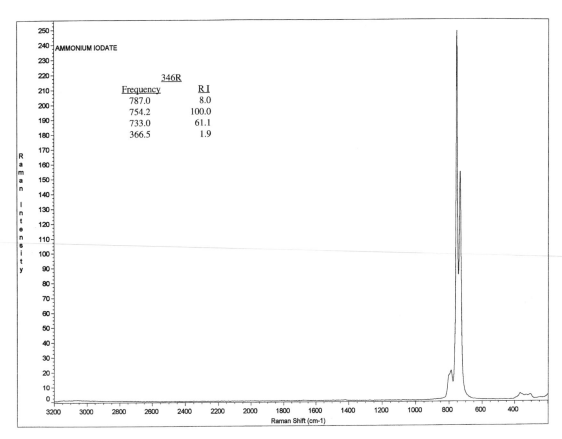

346   Ammonium iodate NH$_4$IO$_3$

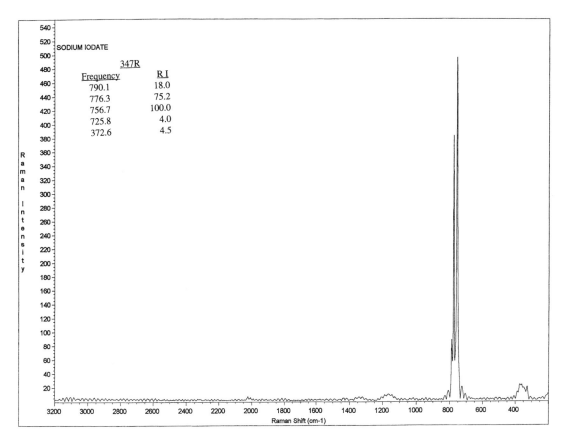

347   Sodium iodate NaIO$_3$

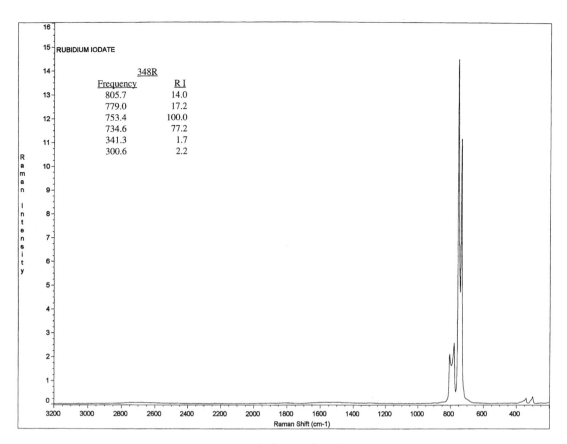

RUBIDIUM IODATE

348R

Frequency	R I
805.7	14.0
779.0	17.2
753.4	100.0
734.6	77.2
341.3	1.7
300.6	2.2

348   Rubidium iodate $RbIO_3$

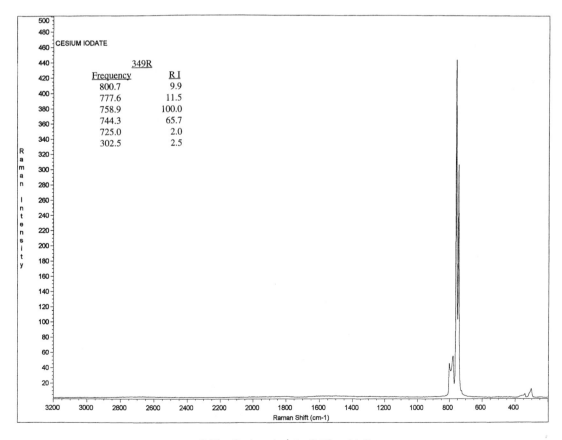

CESIUM IODATE

349R

Frequency	R I
800.7	9.9
777.6	11.5
758.9	100.0
744.3	65.7
725.0	2.0
302.5	2.5

349   Cesium iodate $CsIO_3 \cdot xH_2O$

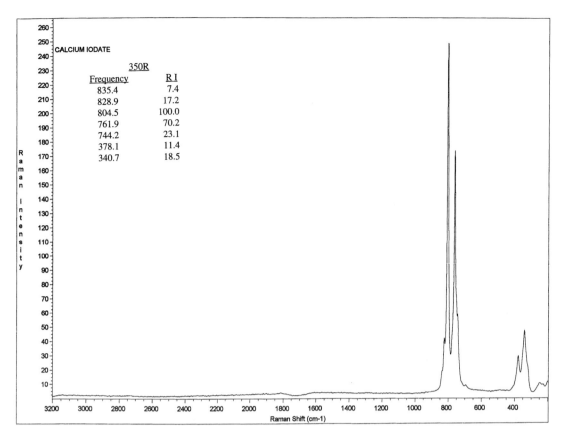

CALCIUM IODATE

350R	
Frequency	R I
835.4	7.4
828.9	17.2
804.5	100.0
761.9	70.2
744.2	23.1
378.1	11.4
340.7	18.5

350   Calcium iodate Ca(IO$_3$)$_2$·6H$_2$O

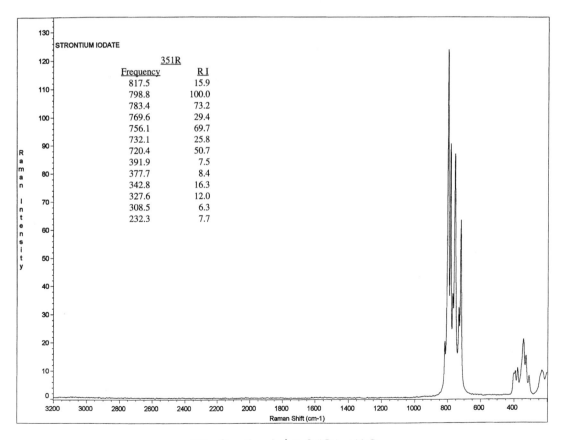

STRONTIUM IODATE

351R	
Frequency	R I
817.5	15.9
798.8	100.0
783.4	73.2
769.6	29.4
756.1	69.7
732.1	25.8
720.4	50.7
391.9	7.5
377.7	8.4
342.8	16.3
327.6	12.0
308.5	6.3
232.3	7.7

351   Strontium iodate Sr(IO$_3$)$_2$·xH$_2$O

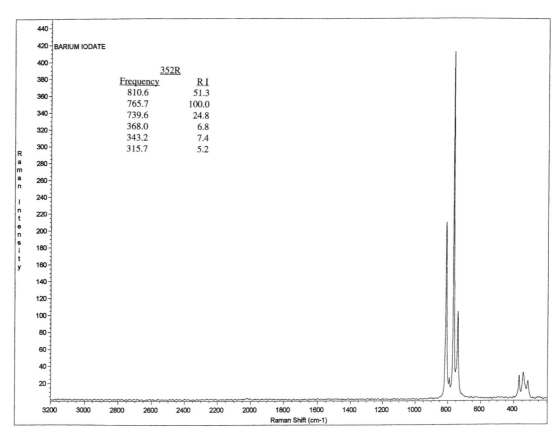

352   Barium iodate $Ba(IO_3)_2 \cdot H_2O$

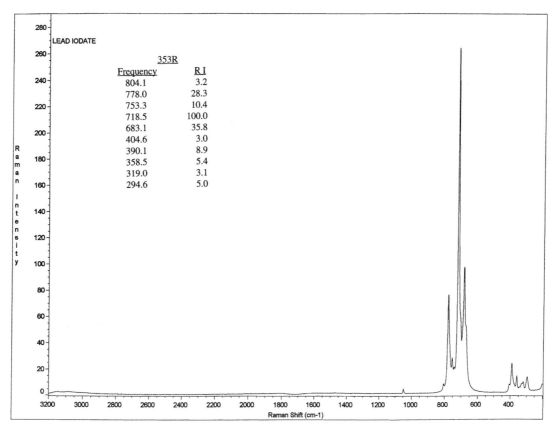

353   Lead iodate $Pb(IO_3)_2$

179

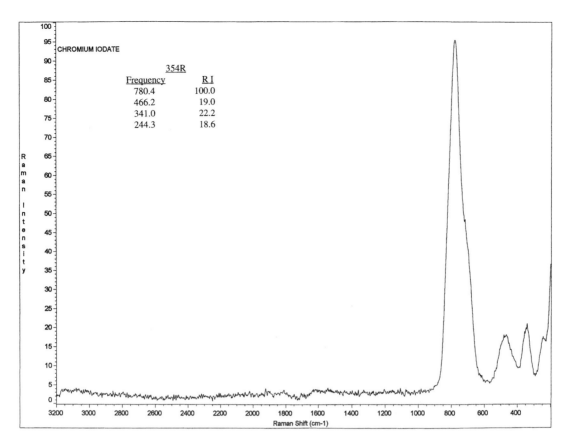

354   Chromium (III) iodate $Cr(IO_3)_3 \cdot xH_2O$

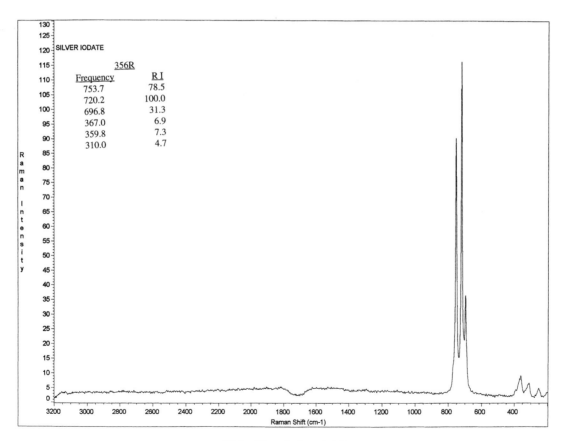

356   Silver iodate $AgIO_3$

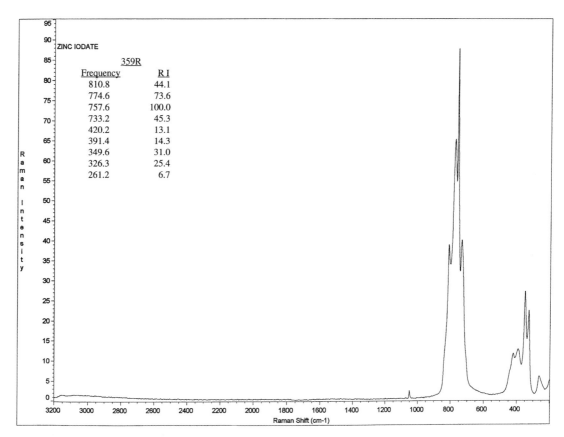

359   Zinc iodate Zn(IO$_3$)$_x$·xH$_2$O

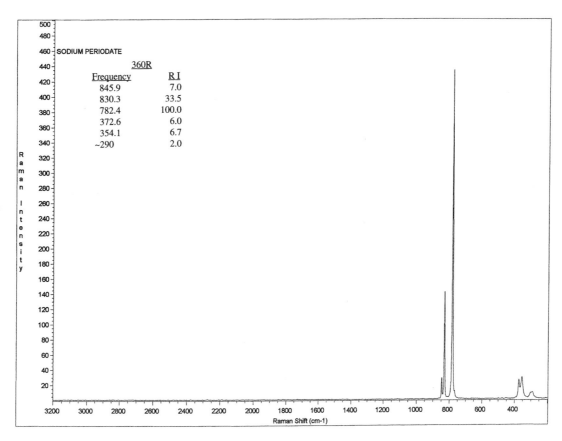

360   Sodium periodate NaIO$_4$

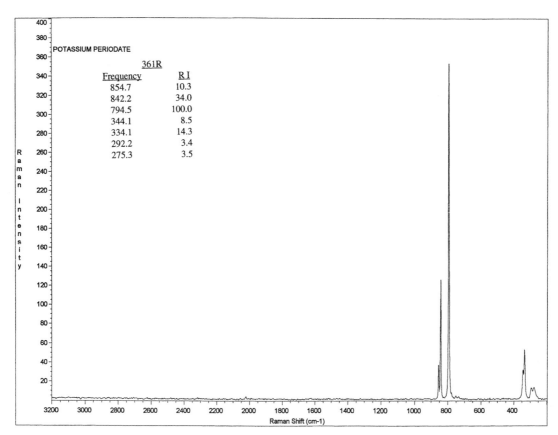

POTASSIUM PERIODATE

361R

Frequency	RI
854.7	10.3
842.2	34.0
794.5	100.0
344.1	8.5
334.1	14.3
292.2	3.4
275.3	3.5

361   Potassium periodate KIO$_4$

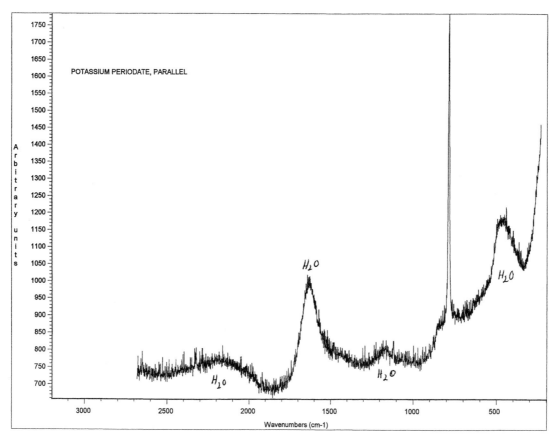

POTASSIUM PERIODATE, PARALLEL

361a   Potassium periodate KIO$_4$ in water solution

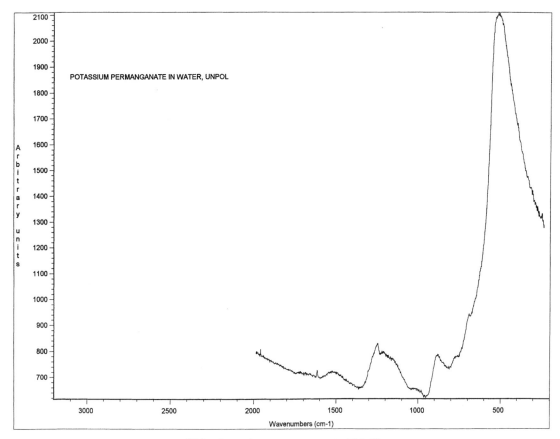

362   Potassium permanganate KMnO$_4$

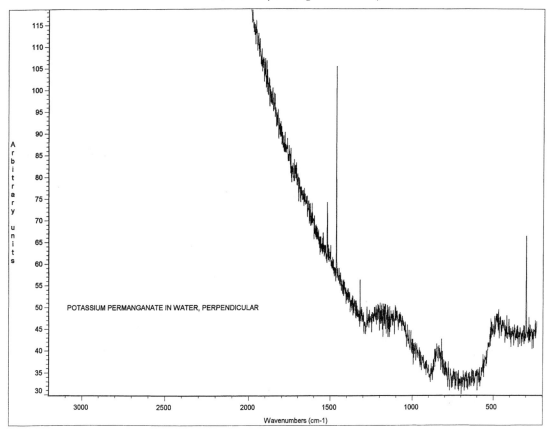

362a   Potassium permanganate KMnO$_4$ in water solution

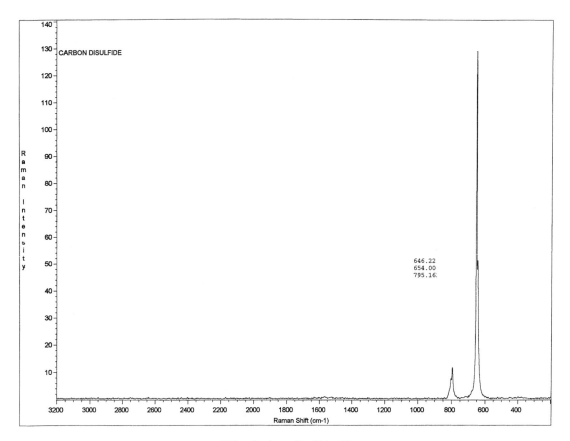

CARBON DISULFIDE

646.22
654.00
795.16

Raman Shift (cm-1)

372   Carbon disulfide CS$_2$

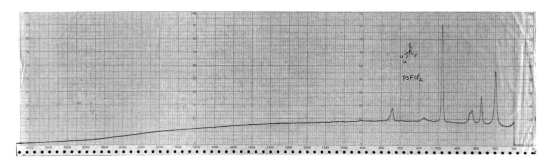

398   Thiophosphoryl dichloride fluoride PCl$_2$FS

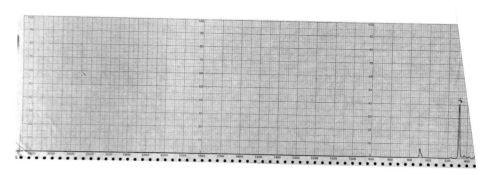

401   Phosphorus pentachloride PCl₅

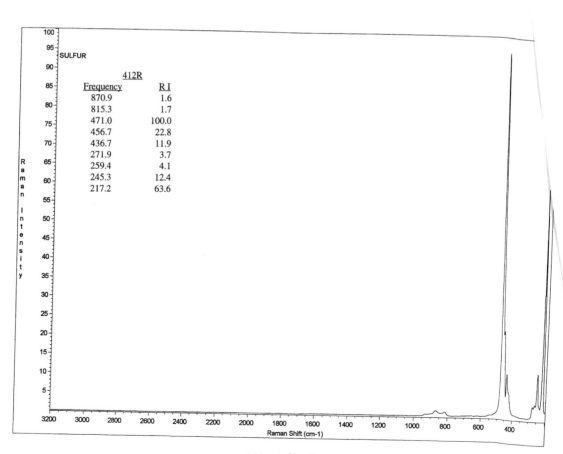

SULFUR

412R	
Frequency	R I
870.9	1.6
815.3	1.7
471.0	100.0
456.7	22.8
436.7	11.9
271.9	3.7
259.4	4.1
245.3	12.4
217.2	63.6

412   Sulfur S₈

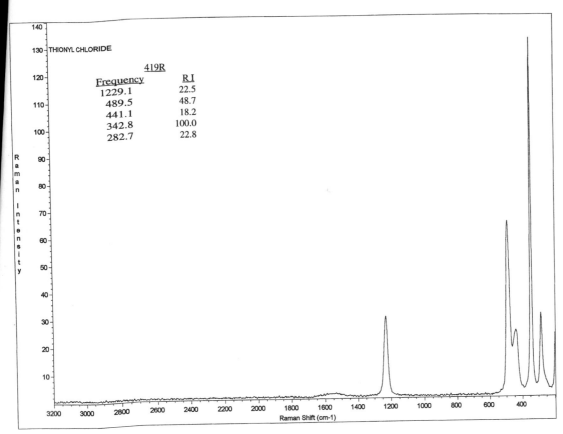

419R	
Frequency	RI
1229.1	22.5
489.5	48.7
441.1	18.2
342.8	100.0
282.7	22.8

THIONYL CHLORIDE

419   Thionyl chloride SCl$_2$O

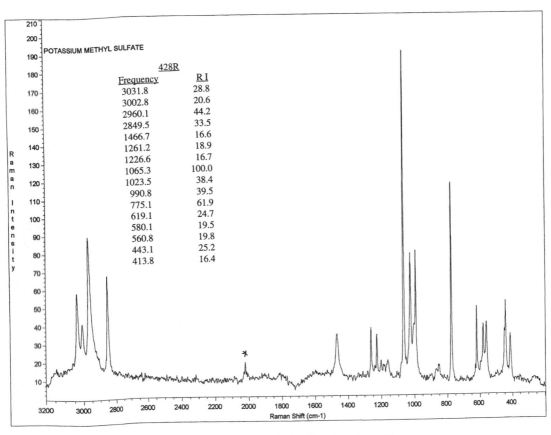

428R	
Frequency	RI
3031.8	28.8
3002.8	20.6
2960.1	44.2
2849.5	33.5
1466.7	16.6
1261.2	18.9
1226.6	16.7
1065.3	100.0
1023.5	38.4
990.8	39.5
775.1	61.9
619.1	24.7
580.1	19.5
560.8	19.8
443.1	25.2
413.8	16.4

POTASSIUM METHYL SULFATE

428   Potassium methyl sulfate CH$_3$OSO$_3$K

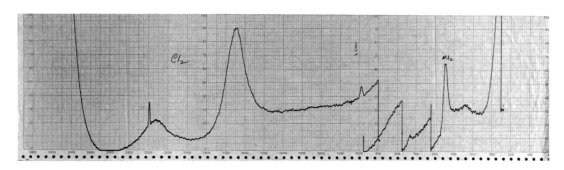

429   Chlorine Cl₂

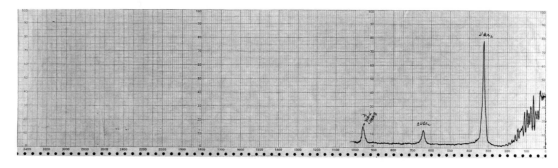

430   Bromine Br₂

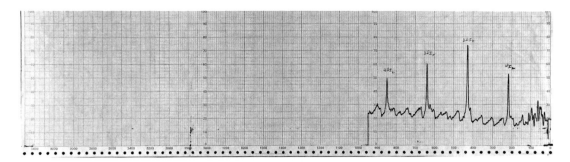

431   Iodine I$_2$

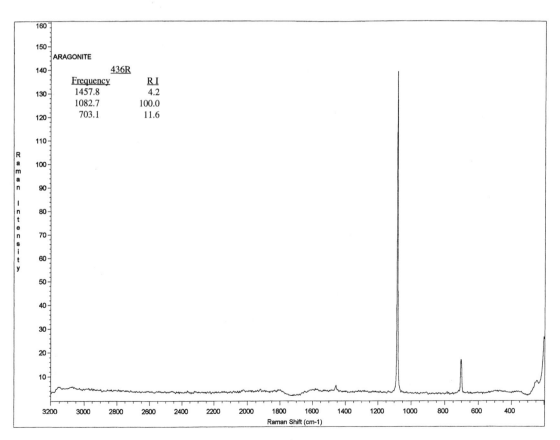

ARAGONITE

	436R	
Frequency		R I
1457.8		4.2
1082.7		100.0
703.1		11.6

436   Aragonite CaCO$_3$

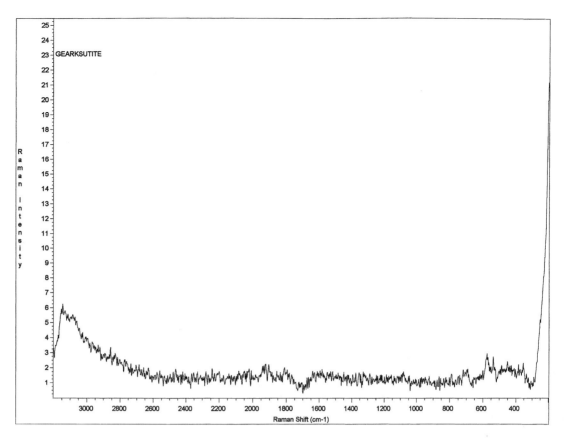

440   Gearksutite CaAlF(OH)

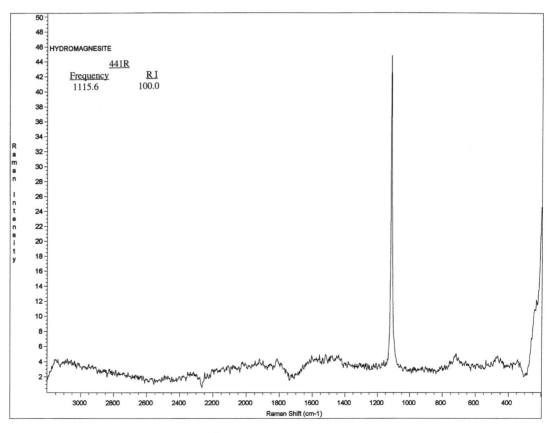

441   Hydromagnesite $3Mg \cdot Mg(OH)_2 \cdot 3H_2O$

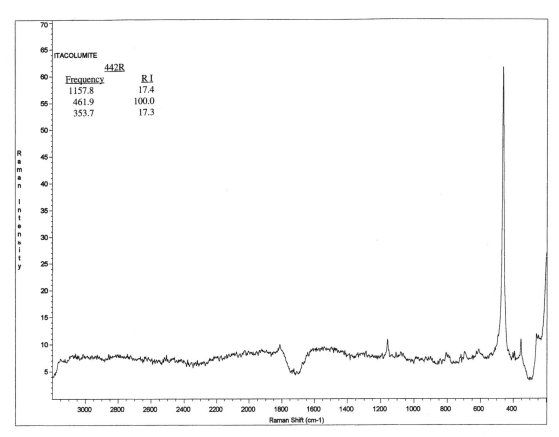

442   Itacolumite

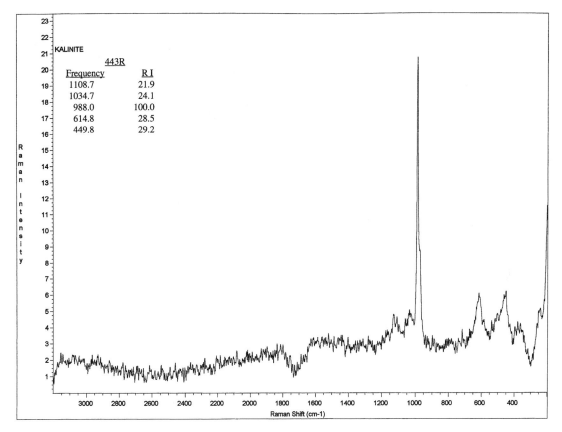

443   Kalinite AlK(SO$_4$)$_2$·12H$_2$O

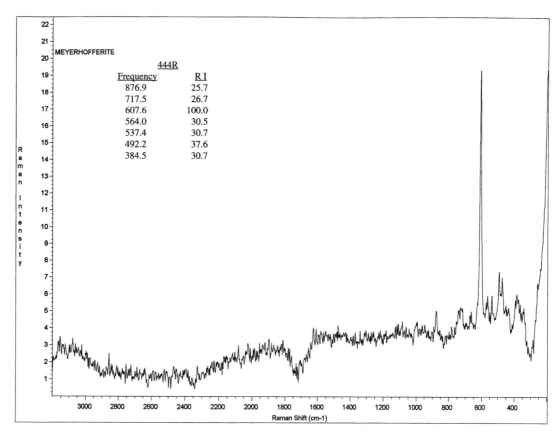

MEYERHOFFERITE	
444R	
Frequency	R I
876.9	25.7
717.5	26.7
607.6	100.0
564.0	30.5
537.4	30.7
492.2	37.6
384.5	30.7

444   Meyerhofferite $2Ca \cdot 3B_2O_3 \cdot 7H_2O$

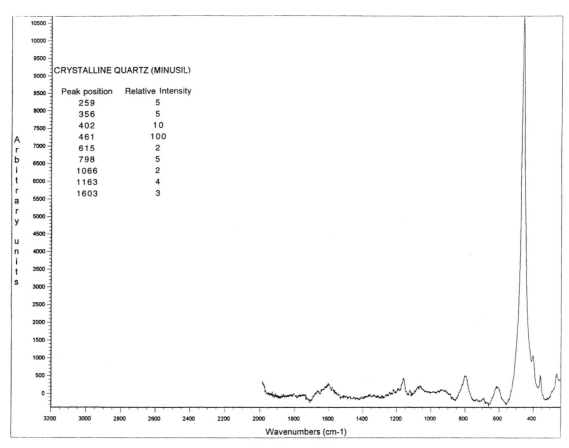

CRYSTALLINE QUARTZ (MINUSIL)	
Peak position	Relative Intensity
259	5
356	5
402	10
461	100
615	2
798	5
1066	2
1163	4
1603	3

445   Quartz SiO_2

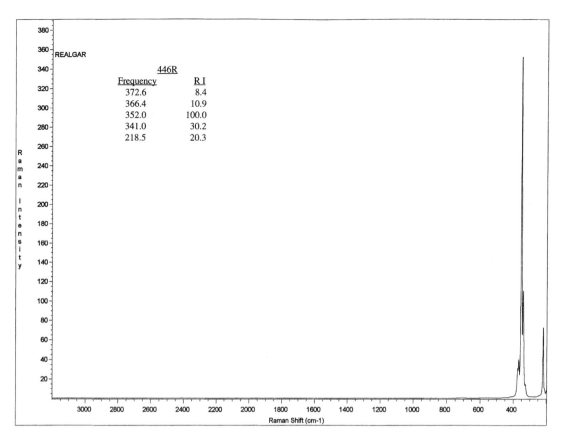

446   Realgar AsS

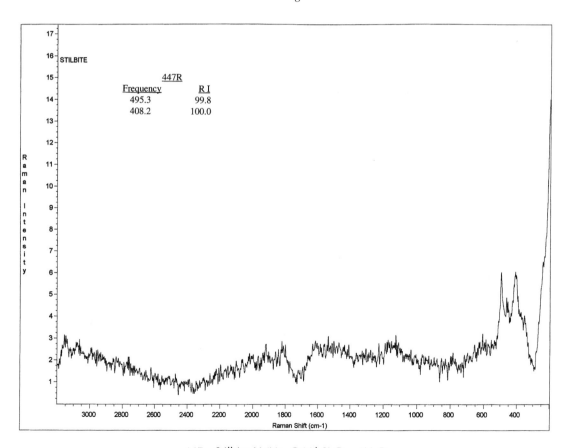

447   Stilbite $H_4(Na_2,Ca)Al_2Si_6O_{18} \cdot 4H_2O$

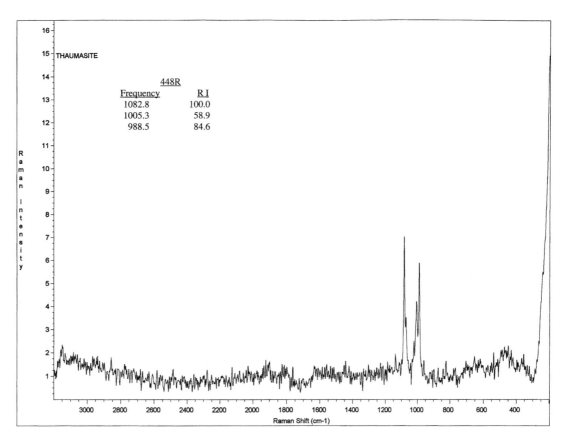

THAUMASITE

448R

Frequency	R I
1082.8	100.0
1005.3	58.9
988.5	84.6

448　Thaumasite $CaSiO_3 \cdot CaCO_3 \cdot CaSO_4 \cdot 15H_2OX$

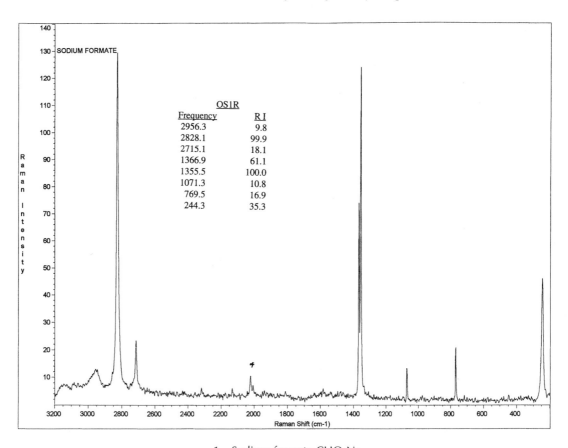

SODIUM FORMATE

OS1R

Frequency	R I
2956.3	9.8
2828.1	99.9
2715.1	18.1
1366.9	61.1
1355.5	100.0
1071.3	10.8
769.5	16.9
244.3	35.3

1　Sodium formate CHO_2Na

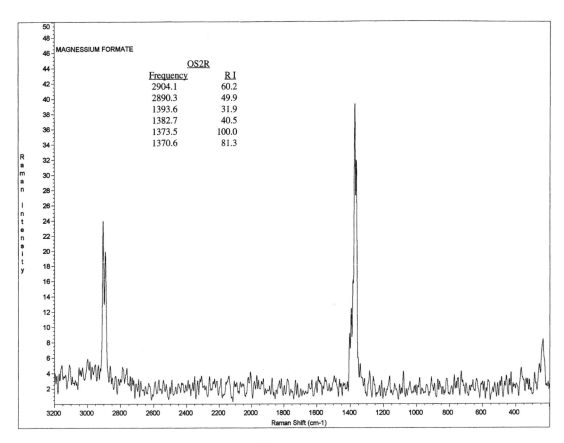

2   Magnesium formate $(CHO_2)_2Mg \cdot xH_2O$

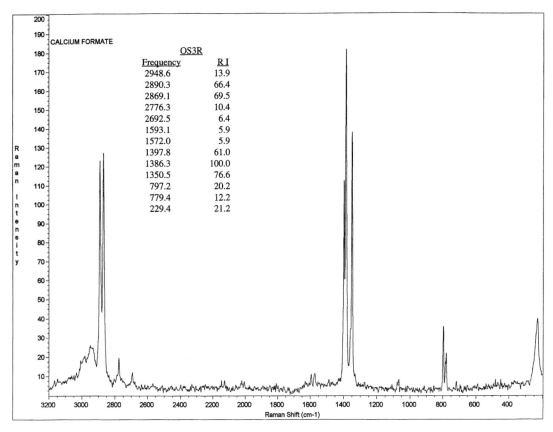

3   Calcium formate $(CHO_2)_2Ca$

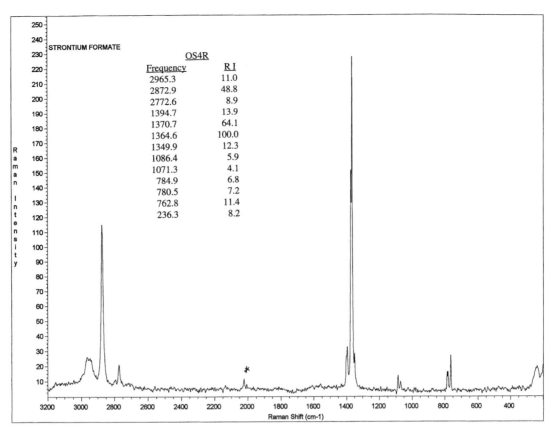

$(CHO_2)_2Sr$

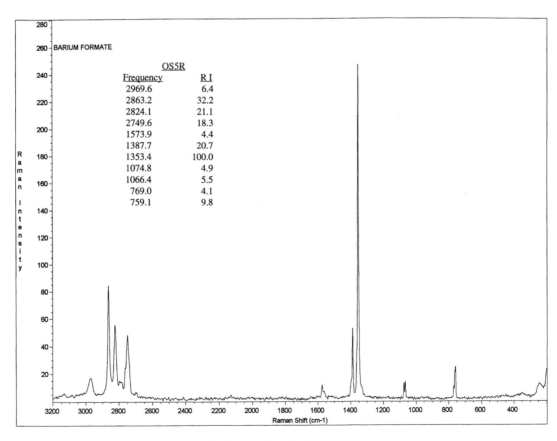

5   Barium formate $(CHO_2)_2Ba$

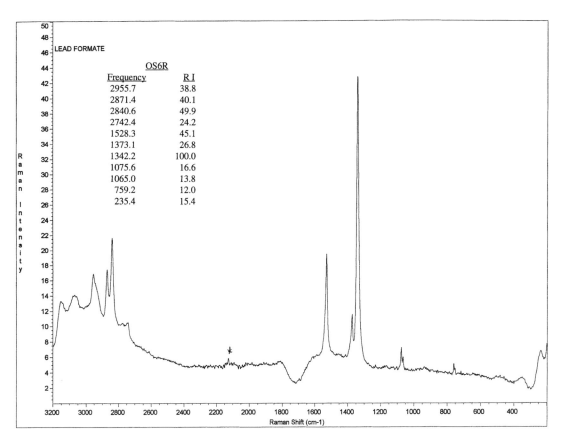

6   Lead formate $(CHO_2)_2Pb$

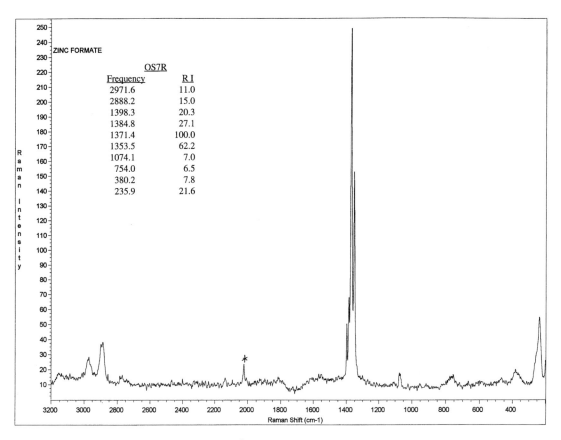

7   Zinc formate $(CHO_2)_2Zn \cdot xH_2O$

196

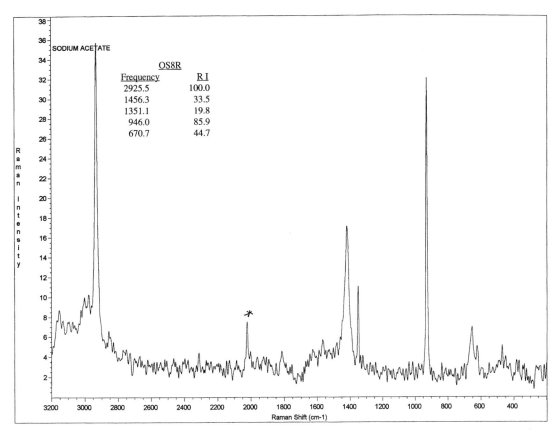

SODIUM ACETATE

OS8R

Frequency	R I
2925.5	100.0
1456.3	33.5
1351.1	19.8
946.0	85.9
670.7	44.7

8   Sodium acetate $(CH_3CO_2)_2Na \cdot xH_2O$

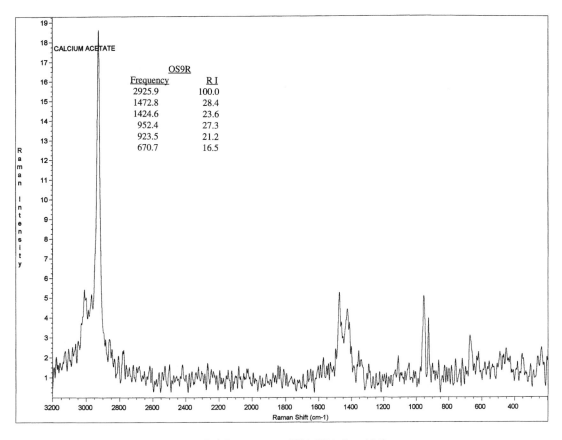

CALCIUM ACETATE

OS9R

Frequency	R I
2925.9	100.0
1472.8	28.4
1424.6	23.6
952.4	27.3
923.5	21.2
670.7	16.5

9   Calcium acetate $(CH_3CO_2)_2Ca \cdot xH_2O$

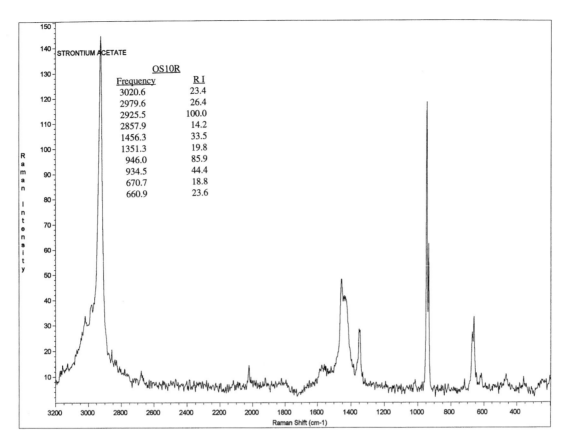

10   Strontium acetate $(CH_3CO_2)_2Sr \cdot xH_2O$

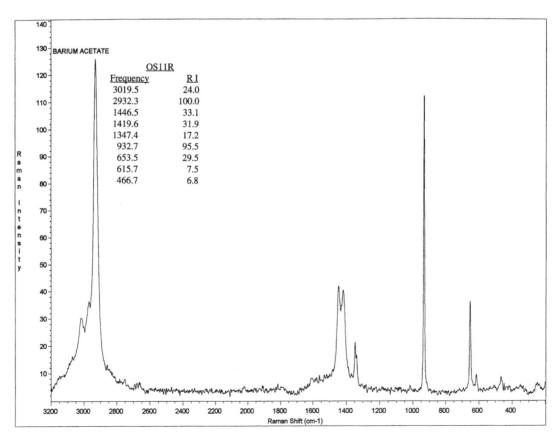

11   Barium acetate $(CH_3CO_2)_2Ba \cdot xH_2O$

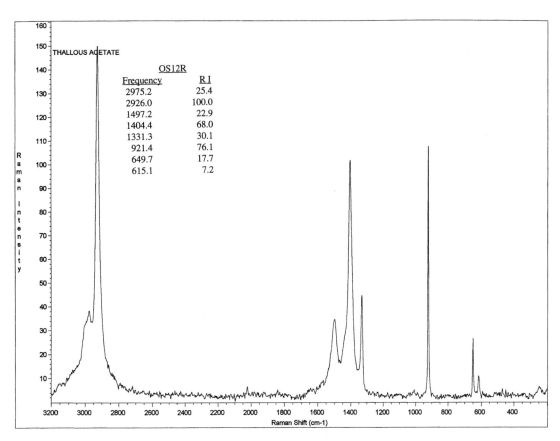

12   Thallous acetate (CH₃CO₂)Tl·xH₂O

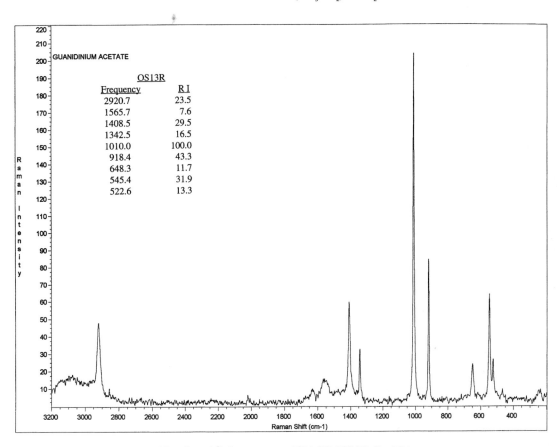

13   Guanidinium acetate (CH₃CO₂)(H₂N)₂C=NH₂

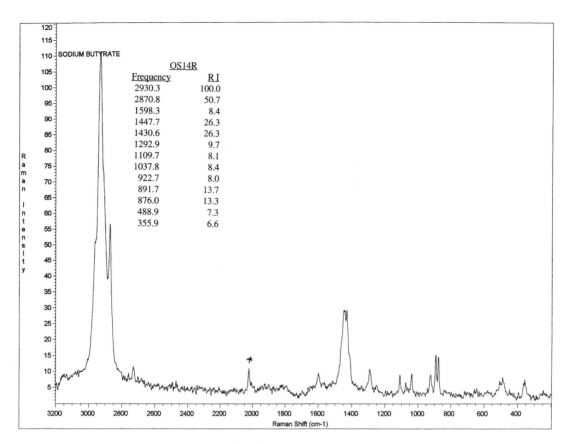

OS14R	
Frequency	R I
2930.3	100.0
2870.8	50.7
1598.3	8.4
1447.7	26.3
1430.6	26.3
1292.9	9.7
1109.7	8.1
1037.8	8.4
922.7	8.0
891.7	13.7
876.0	13.3
488.9	7.3
355.9	6.6

14   Sodium butyrate (n-C$_3$H$_7$CO$_2$)Na

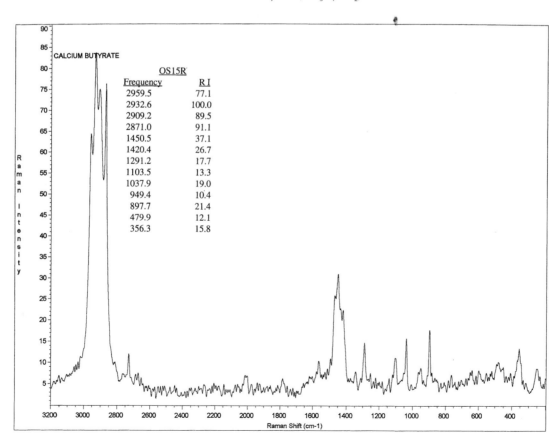

OS15R	
Frequency	R I
2959.5	77.1
2932.6	100.0
2909.2	89.5
2871.0	91.1
1450.5	37.1
1420.4	26.7
1291.2	17.7
1103.5	13.3
1037.9	19.0
949.4	10.4
897.7	21.4
479.9	12.1
356.3	15.8

15   Calcium butyrate (n-C$_3$H$_7$CO$_2$)$_2$Ca

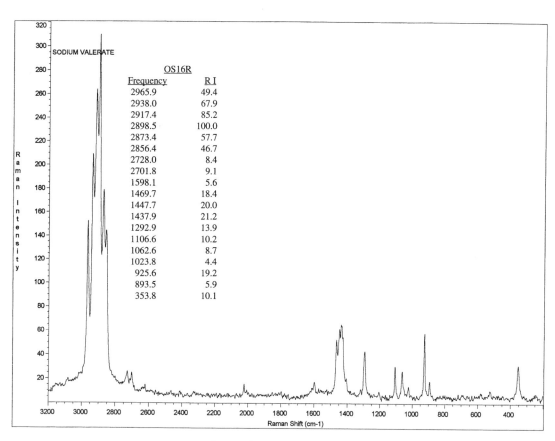

16 Sodium valerate (n-$C_4H_9CO_2$)Na·xH_2O

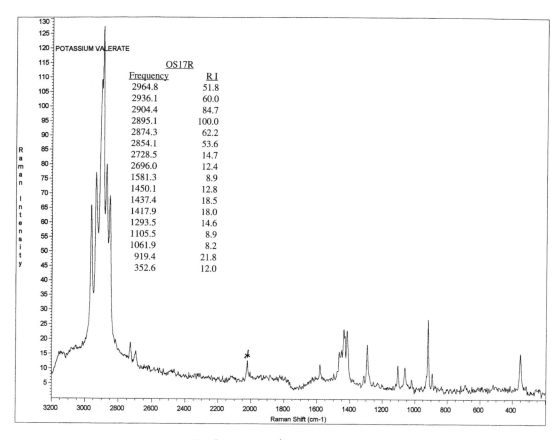

17 Potassium valerate (n-$C_4H_9CO_2$)K

201

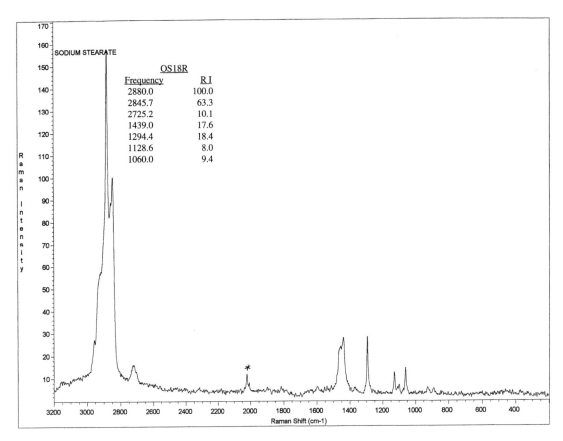

18　Sodium stearate (n-C₁₇H₃₅CO₂)Na

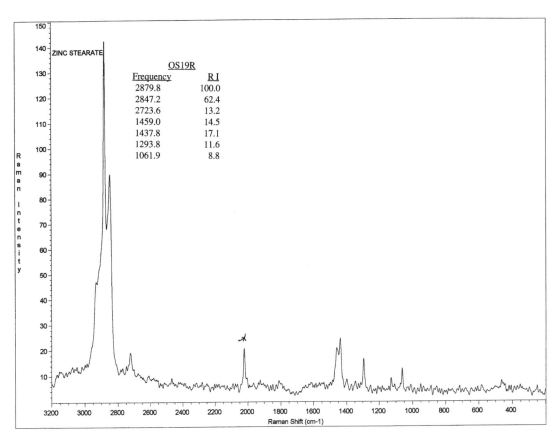

19　Zinc stearate (n-C₁₇H₃₅CO₂)₂Zn

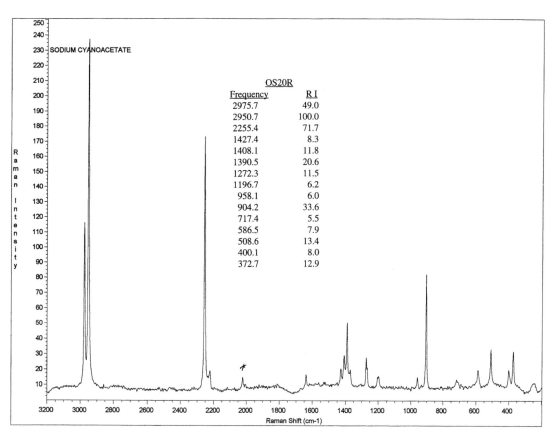

OS20R

Frequency	R I
2975.7	49.0
2950.7	100.0
2255.4	71.7
1427.4	8.3
1408.1	11.8
1390.5	20.6
1272.3	11.5
1196.7	6.2
958.1	6.0
904.2	33.6
717.4	5.5
586.5	7.9
508.6	13.4
400.1	8.0
372.7	12.9

20   Sodium cyanoacetate $(NCCH_2CO_2)Na$

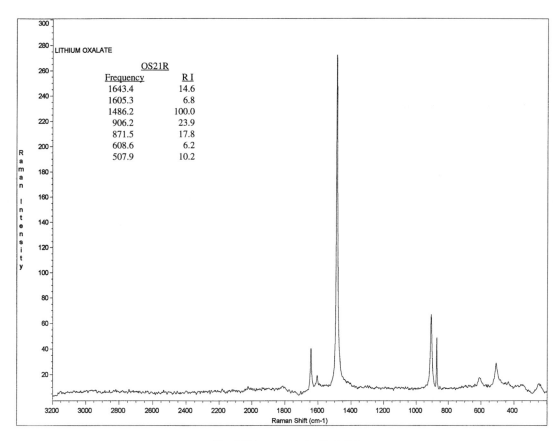

OS21R

Frequency	R I
1643.4	14.6
1605.3	6.8
1486.2	100.0
906.2	23.9
871.5	17.8
608.6	6.2
507.9	10.2

21   Lithium oxalate $(C_2O_4)Li_2$

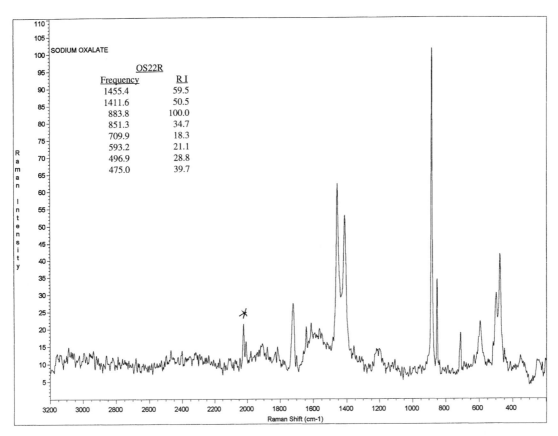

22   Sodium oxalate (C$_2$O$_4$)Na$_2$

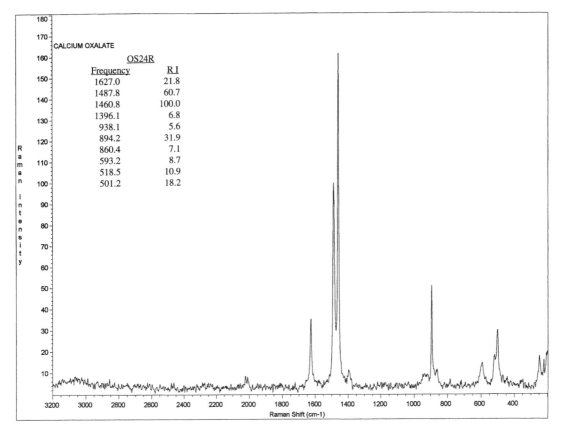

24   Calcium oxalate (C$_2$O$_4$)Ca

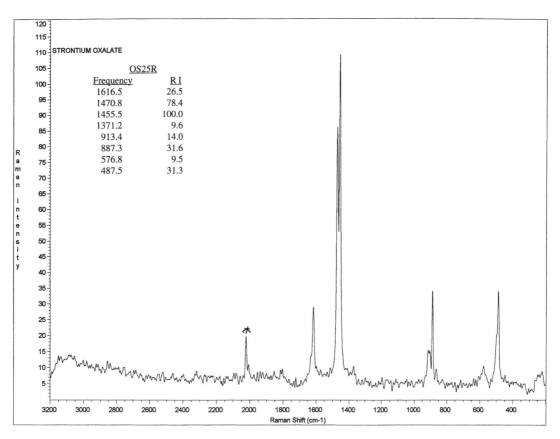

25 Strontium oxalate $(C_2O_4)Sr \cdot xH_2O$

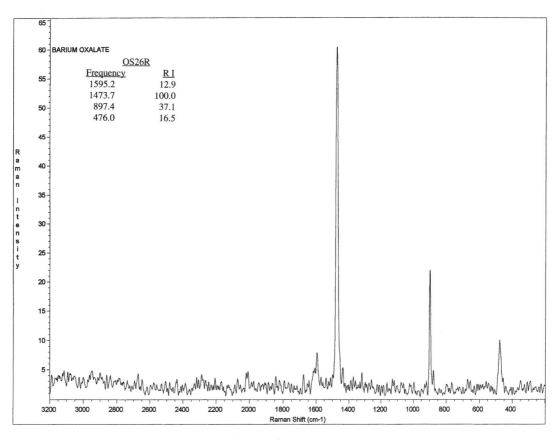

26 Barium oxalate $(C_2O_4)Ba \cdot xH_2O$

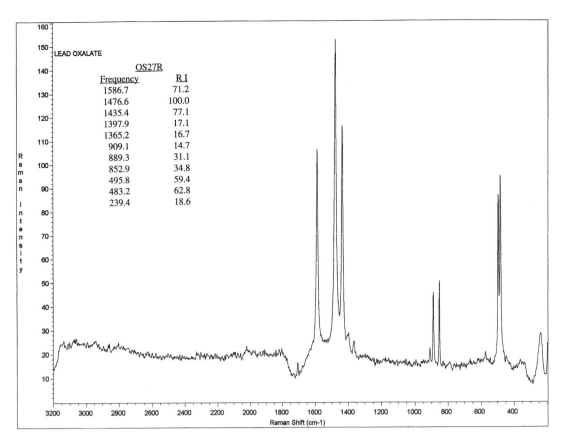

27   Lead oxalate (C$_2$O$_4$)Pb

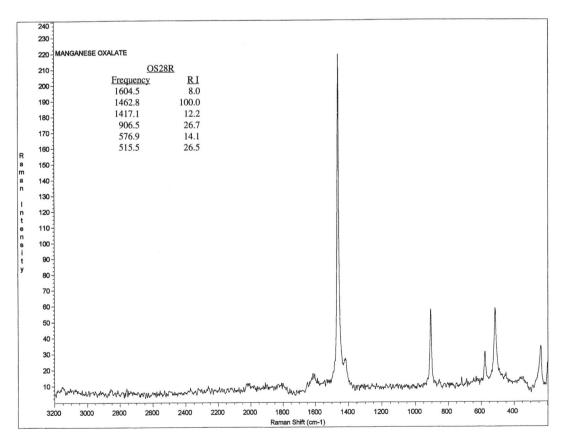

28   Manganese oxalate (C$_2$O$_4$)Mn·xH$_2$O

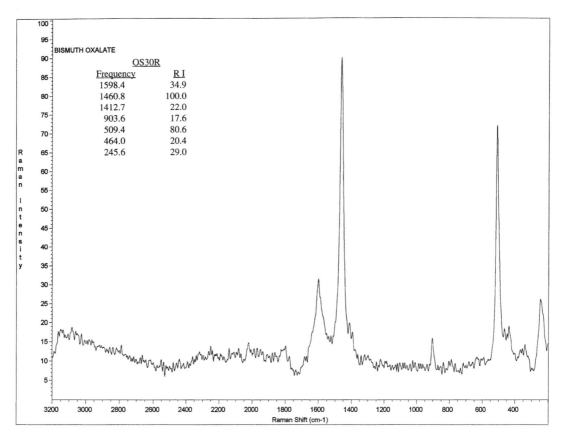

30   Bismuth oxalate $(C_2O_4)Bi \cdot xH_2O$

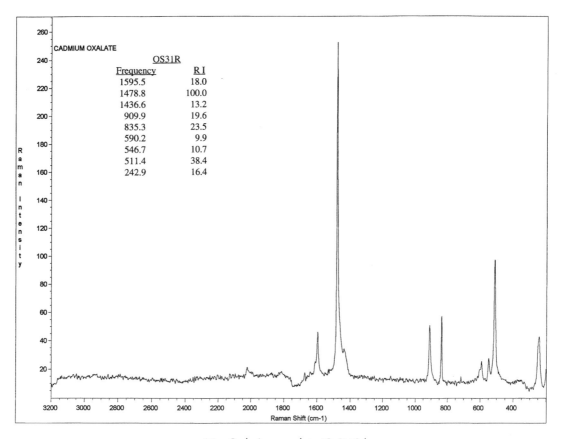

31   Cadmium oxalate $(C_2O_4)Cd$

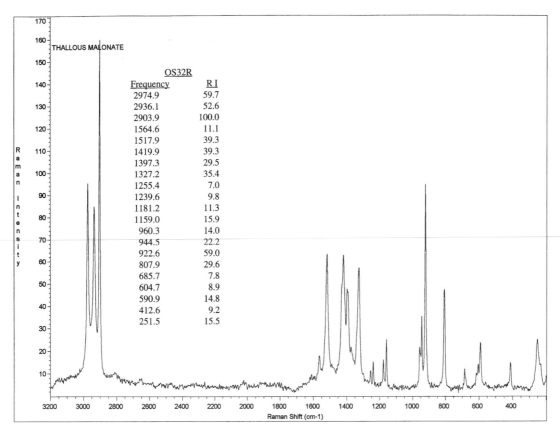

THALLOUS MALONATE

OS32R

Frequency	R I
2974.9	59.7
2936.1	52.6
2903.9	100.0
1564.6	11.1
1517.9	39.3
1419.9	39.3
1397.3	29.5
1327.2	35.4
1255.4	7.0
1239.6	9.8
1181.2	11.3
1159.0	15.9
960.3	14.0
944.5	22.2
922.6	59.0
807.9	29.6
685.7	7.8
604.7	8.9
590.9	14.8
412.6	9.2
251.5	15.5

32　Thallous malonate $(CH_2(CO_2)_2)Tl_2 \cdot xH_2O$

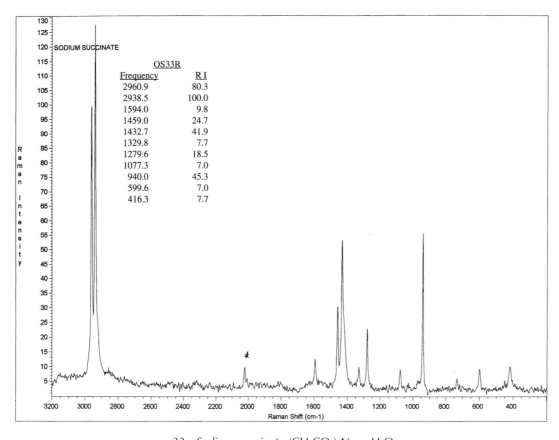

SODIUM SUCCINATE

OS33R

Frequency	R I
2960.9	80.3
2938.5	100.0
1594.0	9.8
1459.0	24.7
1432.7	41.9
1329.8	7.7
1279.6	18.5
1077.3	7.0
940.0	45.3
599.6	7.0
416.3	7.7

33　Sodium succinate $(CH_2CO_2)_2Na_2 \cdot xH_2O$

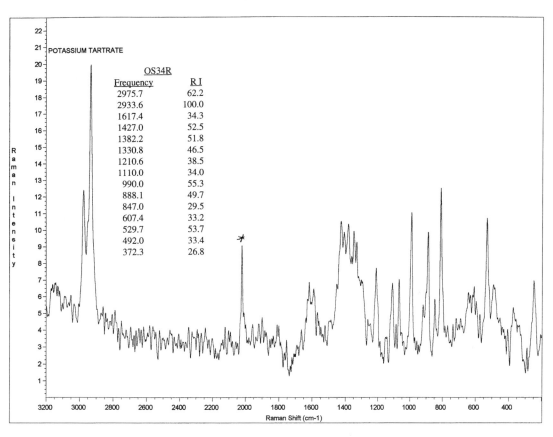

34   Potassium tartrate (CHOHCO$_2$)$_2$K$_2$

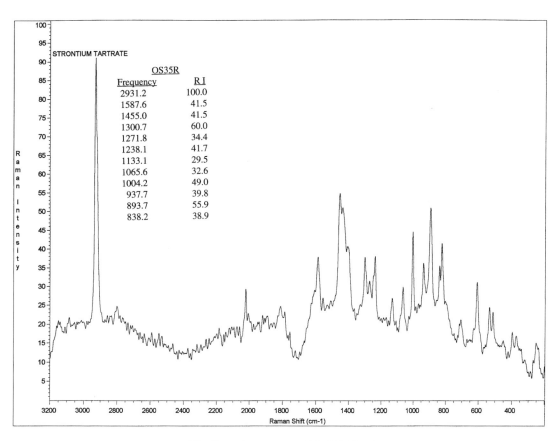

35   Strontium tartrate (CHOHCO$_2$)$_2$Sr

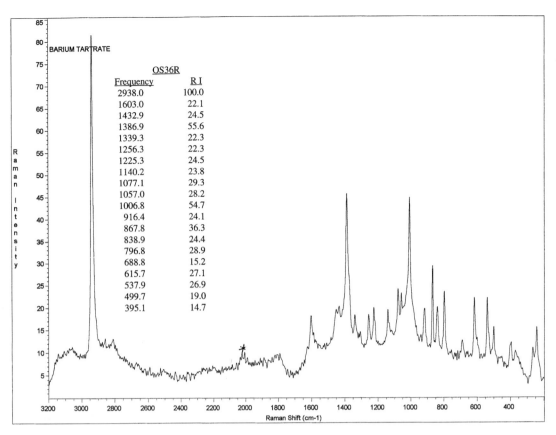

36   Barium tartrate $(CHOHCO_2)_2Ba$

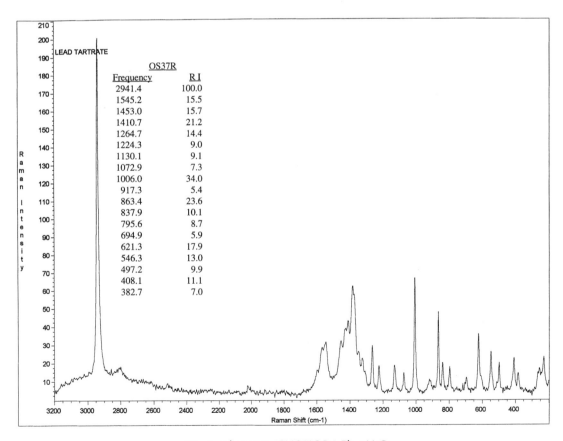

37   Lead tartrate $(CHOHCO_2)_2Pb \cdot xH_2O$

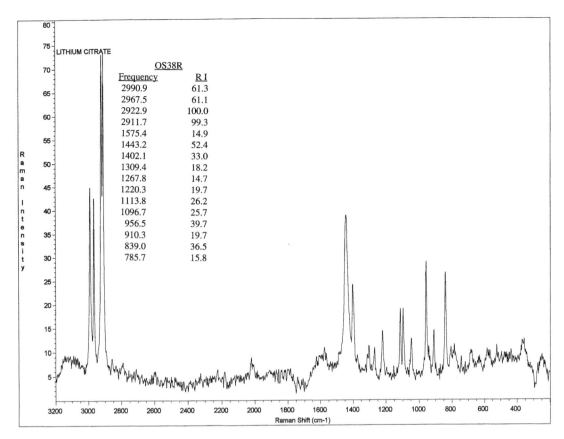

OS38R

Frequency	R I
2990.9	61.3
2967.5	61.1
2922.9	100.0
2911.7	99.3
1575.4	14.9
1443.2	52.4
1402.1	33.0
1309.4	18.2
1267.8	14.7
1220.3	19.7
1113.8	26.2
1096.7	25.7
956.5	39.7
910.3	19.7
839.0	36.5
785.7	15.8

38   Lithium citrate ($(O_2CCH_2)C(OH)(CO_2))Li_3$

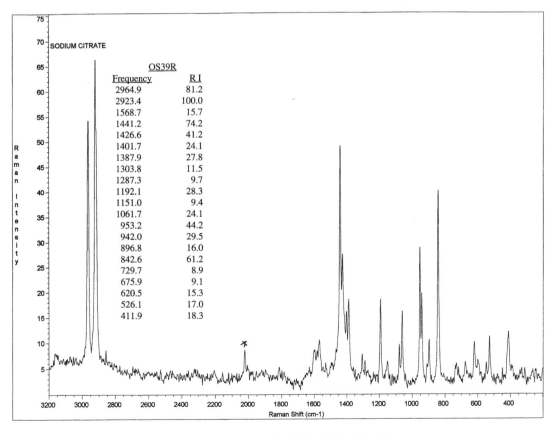

OS39R

Frequency	R I
2964.9	81.2
2923.4	100.0
1568.7	15.7
1441.2	74.2
1426.6	41.2
1401.7	24.1
1387.9	27.8
1303.8	11.5
1287.3	9.7
1192.1	28.3
1151.0	9.4
1061.7	24.1
953.2	44.2
942.0	29.5
896.8	16.0
842.6	61.2
729.7	8.9
675.9	9.1
620.5	15.3
526.1	17.0
411.9	18.3

39   Sodium citrate ($(O_2CCH_2)_2C(OH)(CO_2))Na_3$

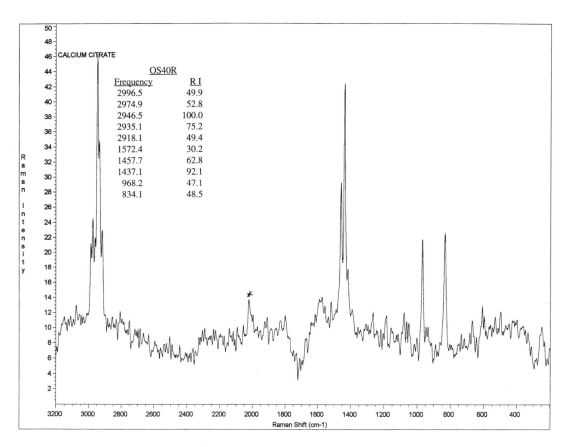

40   Calcium citrate $((O_2CCH_2)_2C(OH)(CO_2))_2Ca_3$

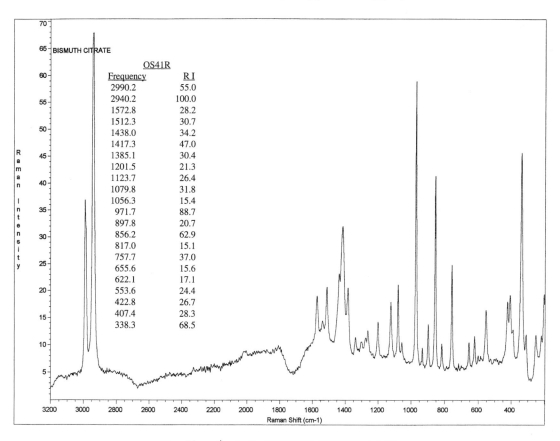

41   Bismuth citrate $((O_2CCH_2)_2C(OH)(CO_2))_2Bi_3$

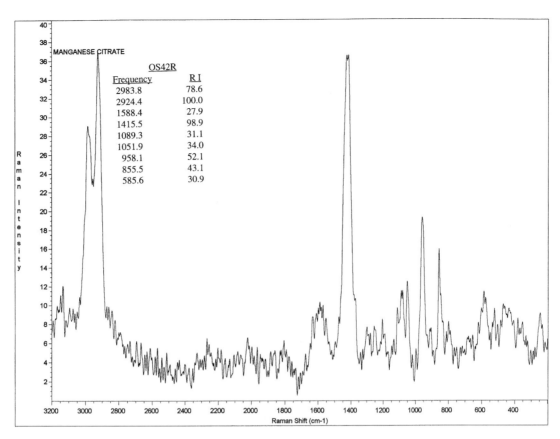

OS42R	
Frequency	R I
2983.8	78.6
2924.4	100.0
1588.4	27.9
1415.5	98.9
1089.3	31.1
1051.9	34.0
958.1	52.1
855.5	43.1
585.6	30.9

42 Manganese citrate $((O_2CCH_2)_2C(OH)(CO_2))_2Mn$

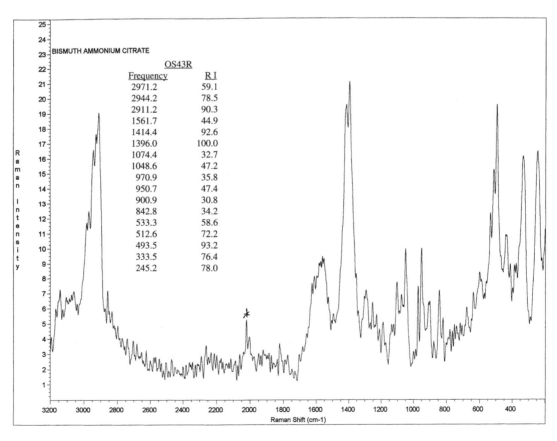

OS43R	
Frequency	R I
2971.2	59.1
2944.2	78.5
2911.2	90.3
1561.7	44.9
1414.4	92.6
1396.0	100.0
1074.4	32.7
1048.6	47.2
970.9	35.8
950.7	47.4
900.9	30.8
842.8	34.2
533.3	58.6
512.6	72.2
493.5	93.2
333.5	76.4
245.2	78.0

43 Bismuth ammonium citrate $((O_2CCH_2)C(OH)(CO_2))BeNH_4$

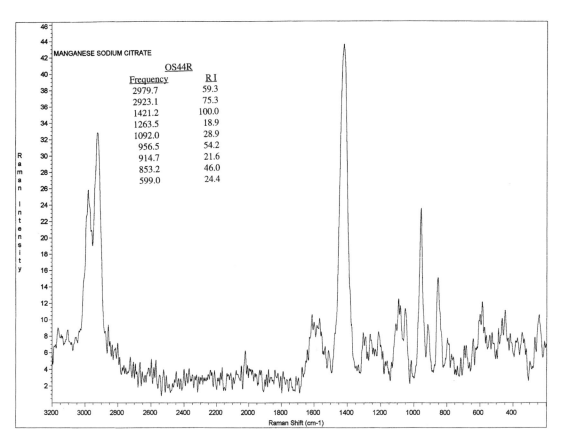

MANGANESE SODIUM CITRATE

OS44R

Frequency	RI
2979.7	59.3
2923.1	75.3
1421.2	100.0
1263.5	18.9
1092.0	28.9
956.5	54.2
914.7	21.6
853.2	46.0
599.0	24.4

44   Manganese sodium citrate $((O_2CCH_2)C(OH)(CO_2))MnNa$

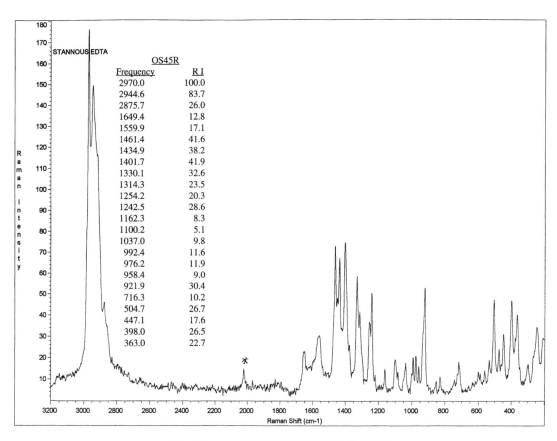

STANNOUS EDTA

OS45R

Frequency	RI
2970.0	100.0
2944.6	83.7
2875.7	26.0
1649.4	12.8
1559.9	17.1
1461.4	41.6
1434.9	38.2
1401.7	41.9
1330.1	32.6
1314.3	23.5
1254.2	20.3
1242.5	28.6
1162.3	8.3
1100.2	5.1
1037.0	9.8
992.4	11.6
976.2	11.9
958.4	9.0
921.9	30.4
716.3	10.2
504.7	26.7
447.1	17.6
398.0	26.5
363.0	22.7

45   Stannous EDTA $[CH_2–N(–CH_2–CO_2)]_2Sn_2$

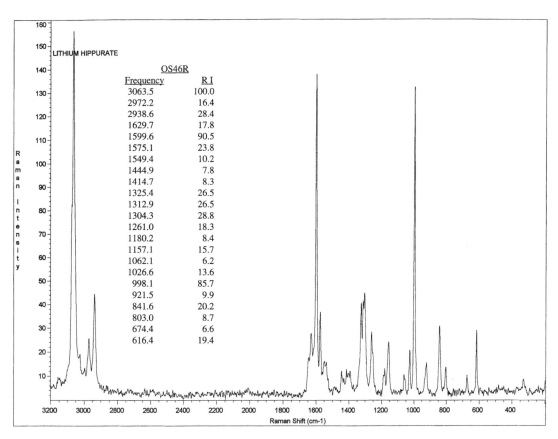

OS46R	
Frequency	R I
3063.5	100.0
2972.2	16.4
2938.6	28.4
1629.7	17.8
1599.6	90.5
1575.1	23.8
1549.4	10.2
1444.9	7.8
1414.7	8.3
1325.4	26.5
1312.9	26.5
1304.3	28.8
1261.0	18.3
1180.2	8.4
1157.1	15.7
1062.1	6.2
1026.6	13.6
998.1	85.7
921.5	9.9
841.6	20.2
803.0	8.7
674.4	6.6
616.4	19.4

46   Lithium hippurate (C$_6$H$_5$C(=O)NHCH$_2$CO$_2$)Li

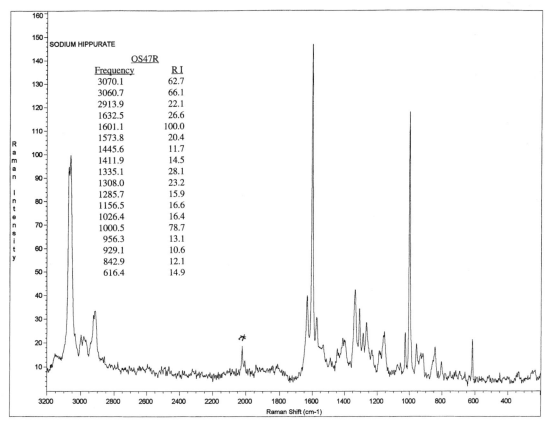

OS47R	
Frequency	R I
3070.1	62.7
3060.7	66.1
2913.9	22.1
1632.5	26.6
1601.1	100.0
1573.8	20.4
1445.6	11.7
1411.9	14.5
1335.1	28.1
1308.0	23.2
1285.7	15.9
1156.5	16.6
1026.4	16.4
1000.5	78.7
956.3	13.1
929.1	10.6
842.9	12.1
616.4	14.9

47   Sodium hippurate (C$_6$H$_5$C(=O)NHCH$_2$CO$_2$)Na

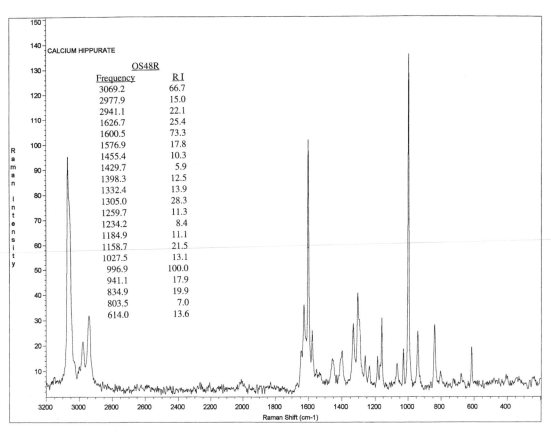

CALCIUM HIPPURATE	
OS48R	
Frequency	R I
3069.2	66.7
2977.9	15.0
2941.1	22.1
1626.7	25.4
1600.5	73.3
1576.9	17.8
1455.4	10.3
1429.7	5.9
1398.3	12.5
1332.4	13.9
1305.0	28.3
1259.7	11.3
1234.2	8.4
1184.9	11.1
1158.7	21.5
1027.5	13.1
996.9	100.0
941.1	17.9
834.9	19.9
803.5	7.0
614.0	13.6

48   Calcium hippurate [(($C_6H_5C(=O)NHCH_2CO_2$)]Ca

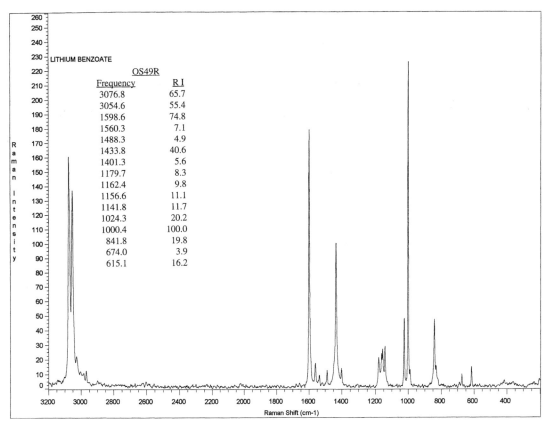

LITHIUM BENZOATE	
OS49R	
Frequency	R I
3076.8	65.7
3054.6	55.4
1598.6	74.8
1560.3	7.1
1488.3	4.9
1433.8	40.6
1401.3	5.6
1179.7	8.3
1162.4	9.8
1156.6	11.1
1141.8	11.7
1024.3	20.2
1000.4	100.0
841.8	19.8
674.0	3.9
615.1	16.2

49   Lithium benzoate ($C_6H_5CO_2$)Li

216

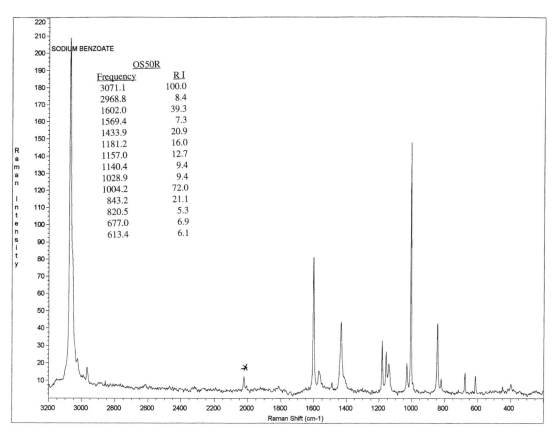

OS50R	
Frequency	R I
3071.1	100.0
2968.8	8.4
1602.0	39.3
1569.4	7.3
1433.9	20.9
1181.2	16.0
1157.0	12.7
1140.4	9.4
1028.9	9.4
1004.2	72.0
843.2	21.1
820.5	5.3
677.0	6.9
613.4	6.1

50   Sodium benzoate (C$_6$H$_5$CO$_2$)Na

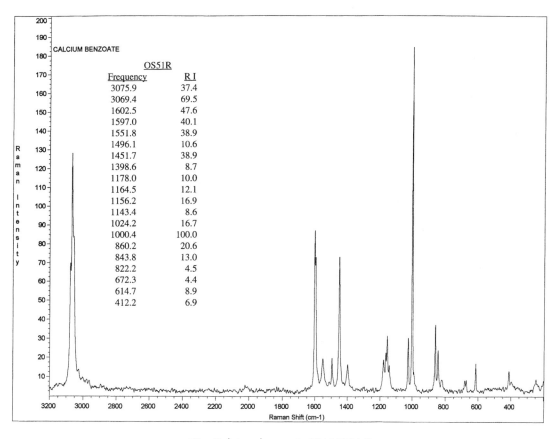

OS51R	
Frequency	R I
3075.9	37.4
3069.4	69.5
1602.5	47.6
1597.0	40.1
1551.8	38.9
1496.1	10.6
1451.7	38.9
1398.6	8.7
1178.0	10.0
1164.5	12.1
1156.2	16.9
1143.4	8.6
1024.2	16.7
1000.4	100.0
860.2	20.6
843.8	13.0
822.2	4.5
672.3	4.4
614.7	8.9
412.2	6.9

51   Calcium benzoate (C$_6$H$_5$CO$_2$)$_2$Ca

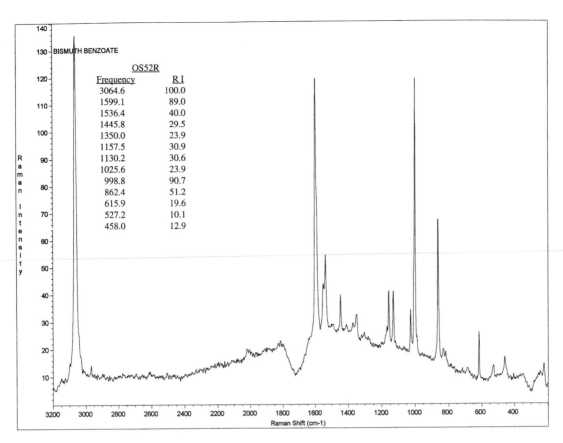

52   Bismuth benzoate $(C_6H_5CO_2)_2Bi$

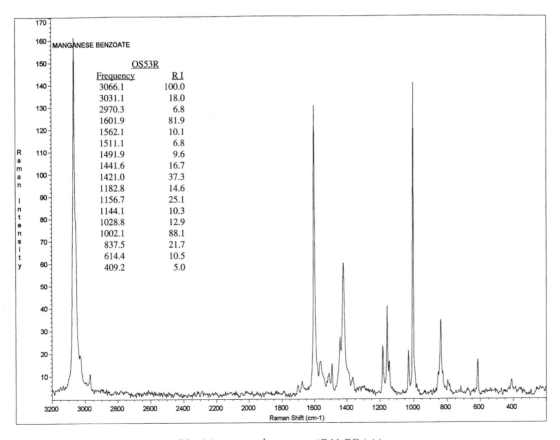

53   Manganese benzoate $(C_6H_5CO_2)_2Mn$

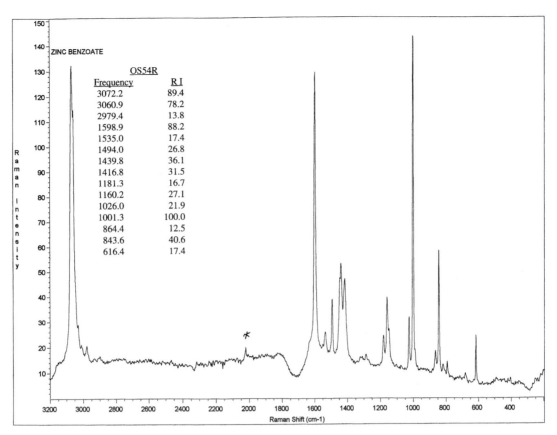

	OS54R	
Frequency		R I
3072.2		89.4
3060.9		78.2
2979.4		13.8
1598.9		88.2
1535.0		17.4
1494.0		26.8
1439.8		36.1
1416.8		31.5
1181.3		16.7
1160.2		27.1
1026.0		21.9
1001.3		100.0
864.4		12.5
843.6		40.6
616.4		17.4

54   Zinc benzoate $(C_6H_5CO_2)_2Zn$

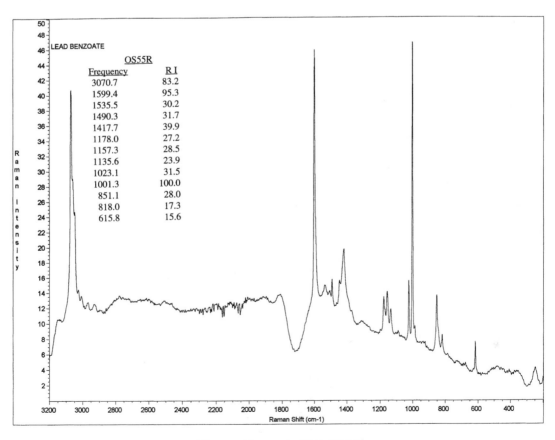

	OS55R	
Frequency		R I
3070.7		83.2
1599.4		95.3
1535.5		30.2
1490.3		31.7
1417.7		39.9
1178.0		27.2
1157.3		28.5
1135.6		23.9
1023.1		31.5
1001.3		100.0
851.1		28.0
818.0		17.3
615.8		15.6

55   Lead benzoate $(C_6H_5CO_2)_2Pb$

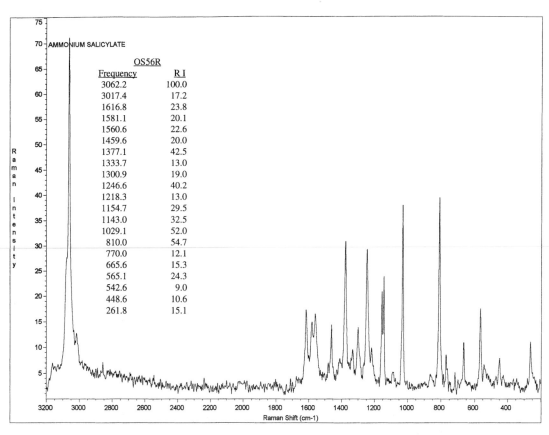

56   Ammonium salicylate (o-(OH)C$_6$H$_4$CO$_2$)NH$_4$

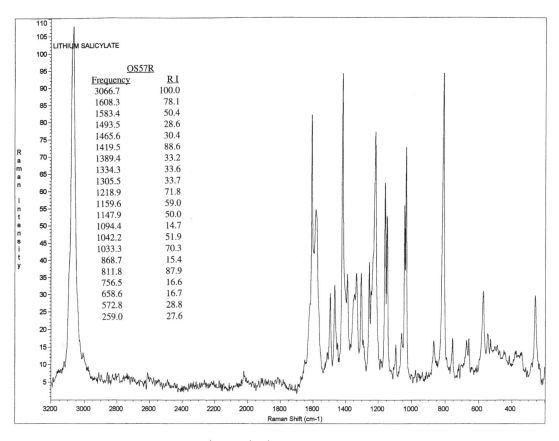

57   Lithium salicylate (o-(OH)C$_6$H$_4$CO$_2$)Li

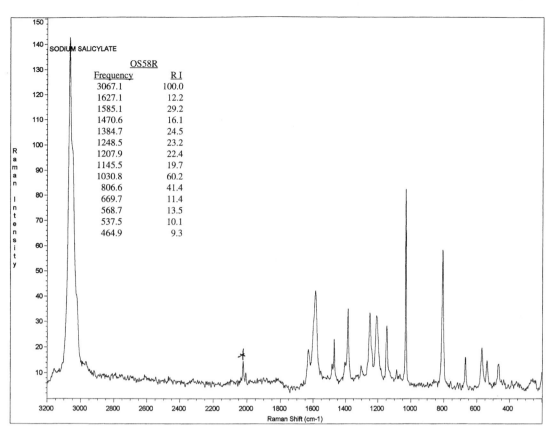

OS58R	
Frequency	R I
3067.1	100.0
1627.1	12.2
1585.1	29.2
1470.6	16.1
1384.7	24.5
1248.5	23.2
1207.9	22.4
1145.5	19.7
1030.8	60.2
806.6	41.4
669.7	11.4
568.7	13.5
537.5	10.1
464.9	9.3

58   Sodium salicylate (o-(OH)C$_6$H$_4$CO$_2$)Na

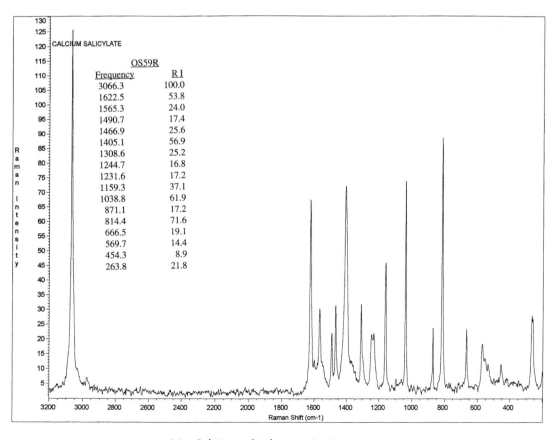

OS59R	
Frequency	R I
3066.3	100.0
1622.5	53.8
1565.3	24.0
1490.7	17.4
1466.9	25.6
1405.1	56.9
1308.6	25.2
1244.7	16.8
1231.6	17.2
1159.3	37.1
1038.8	61.9
871.1	17.2
814.4	71.6
666.5	19.1
569.7	14.4
454.3	8.9
263.8	21.8

59   Calcium salicylate (o-(OH)C$_6$H$_4$CO$_2$)$_2$Ca

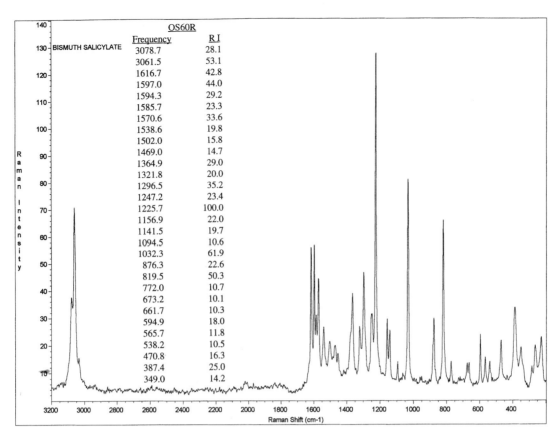

OS60R

Frequency	RI
3078.7	28.1
3061.5	53.1
1616.7	42.8
1597.0	44.0
1594.3	29.2
1585.7	23.3
1570.6	33.6
1538.6	19.8
1502.0	15.8
1469.0	14.7
1364.9	29.0
1321.8	20.0
1296.5	35.2
1247.2	23.4
1225.7	100.0
1156.9	22.0
1141.5	19.7
1094.5	10.6
1032.3	61.9
876.3	22.6
819.5	50.3
772.0	10.7
673.2	10.1
661.7	10.3
594.9	18.0
565.7	11.8
538.2	10.5
470.8	16.3
387.4	25.0
349.0	14.2

60   Bismuth salicylate (o-(OH)C$_6$H$_4$CO$_2$)$_2$Bi

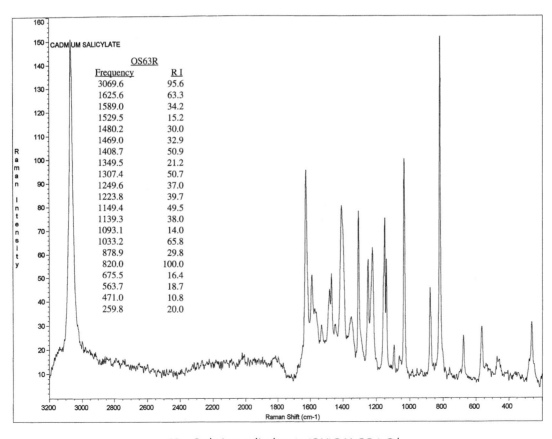

OS63R

Frequency	RI
3069.6	95.6
1625.6	63.3
1589.0	34.2
1529.5	15.2
1480.2	30.0
1469.0	32.9
1408.7	50.9
1349.5	21.2
1307.4	50.7
1249.6	37.0
1223.8	39.7
1149.4	49.5
1139.3	38.0
1093.1	14.0
1033.2	65.8
878.9	29.8
820.0	100.0
675.5	16.4
563.7	18.7
471.0	10.8
259.8	20.0

63   Cadmium salicylate (o-(OH)C$_6$H$_4$CO$_2$)$_2$Cd

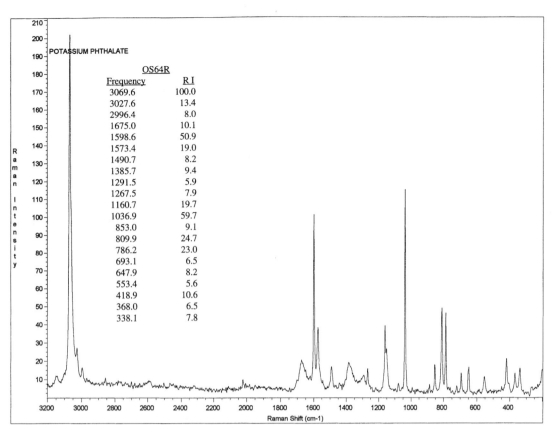

OS64R	
Frequency	R I
3069.6	100.0
3027.6	13.4
2996.4	8.0
1675.0	10.1
1598.6	50.9
1573.4	19.0
1490.7	8.2
1385.7	9.4
1291.5	5.9
1267.5	7.9
1160.7	19.7
1036.9	59.7
853.0	9.1
809.9	24.7
786.2	23.0
693.1	6.5
647.9	8.2
553.4	5.6
418.9	10.6
368.0	6.5
338.1	7.8

64　Potassium phthalate [o-$C_6H_4(CO_2)_2$]K_2

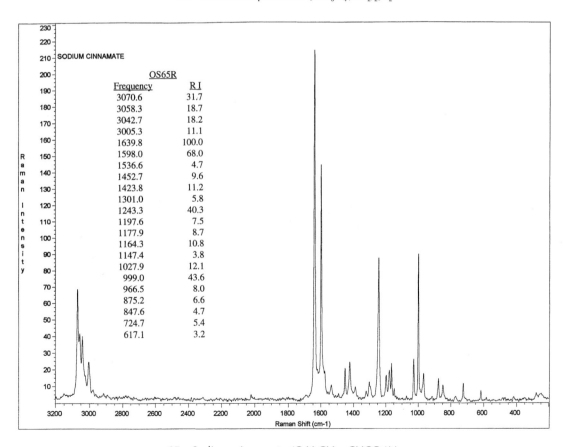

OS65R	
Frequency	R I
3070.6	31.7
3058.3	18.7
3042.7	18.2
3005.3	11.1
1639.8	100.0
1598.0	68.0
1536.6	4.7
1452.7	9.6
1423.8	11.2
1301.0	5.8
1243.3	40.3
1197.6	7.5
1177.9	8.7
1164.3	10.8
1147.4	3.8
1027.9	12.1
999.0	43.6
966.5	8.0
875.2	6.6
847.6	4.7
724.7	5.4
617.1	3.2

65　Sodium cinnamate ($C_6H_5CH{=}CHCO_2$)Na

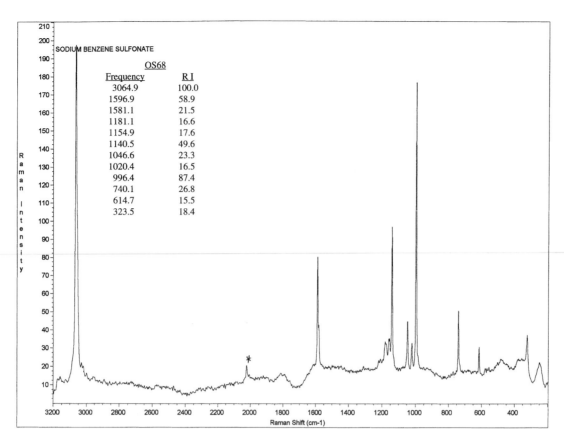

SODIUM BENZENE SULFONATE

OS68

Frequency	R I
3064.9	100.0
1596.9	58.9
1581.1	21.5
1181.1	16.6
1154.9	17.6
1140.5	49.6
1046.6	23.3
1020.4	16.5
996.4	87.4
740.1	26.8
614.7	15.5
323.5	18.4

68   Sodium benzenesulfonate $(C_6H_5SO_3)Na$

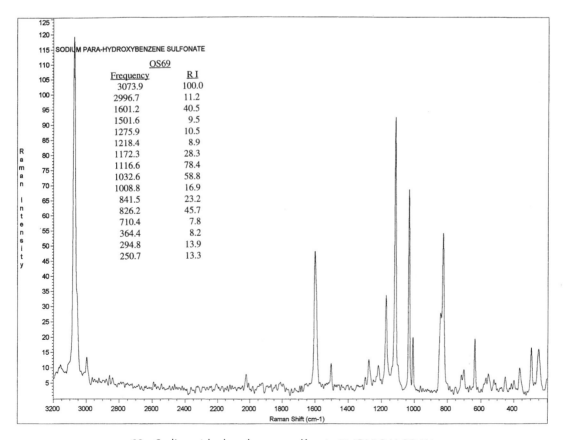

SODIUM PARA-HYDROXYBENZENE SULFONATE

OS69

Frequency	R I
3073.9	100.0
2996.7	11.2
1601.2	40.5
1501.6	9.5
1275.9	10.5
1218.4	8.9
1172.3	28.3
1116.6	78.4
1032.6	58.8
1008.8	16.9
841.5	23.2
826.2	45.7
710.4	7.8
364.4	8.2
294.8	13.9
250.7	13.3

69   Sodium 4-hydroxybenzenesulfonate $(4\text{-}(OH)C_6H_4SO_3)Na$